Abdessalem BADRI

CRYSTALLOGRAPHY for Preparatory Classes

Abdessalem BADRI

CRYSTALLOGRAPHY for Preparatory Classes

Basics; X-ray diffraction; Stacks; Metallic, ionic and covalent crystals

ScienciaScripts

Imprint
Any brand names and product names mentioned in this book are subject to trademark, brand or patent protection and are trademarks or registered trademarks of their respective holders. The use of brand names, product names, common names, trade names, product descriptions etc. even without a particular marking in this work is in no way to be construed to mean that such names may be regarded as unrestricted in respect of trademark and brand protection legislation and could thus be used by anyone.

Cover image: www.ingimage.com

This book is a translation from the original published under ISBN 978-620-6-70629-8.

Publisher:
Sciencia Scripts
is a trademark of
Dodo Books Indian Ocean Ltd. and OmniScriptum S.R.L publishing group

120 High Road, East Finchley, London, N2 9ED, United Kingdom
Str. Armeneasca 28/1, office 1, Chisinau MD-2012, Republic of Moldova, Europe
Printed at: see last page
ISBN: 978-620-7-71124-6

Contents

CHAPTER 1

Introduction and basic concepts of crystallography

CHAPTER1

Introduction and basic concepts of crystallography

I. Physical state of matter

States of matter

gas is fluid, expansive and disorderly
liquid is fluid, condensed and disordered
solid is condensed, rigid and usually ordered

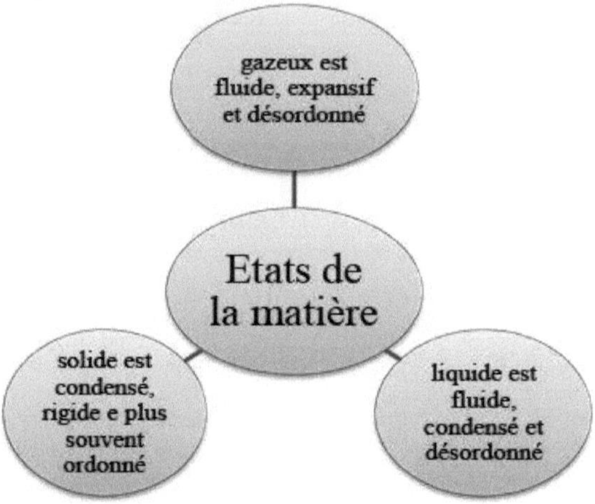

The solid state is subdivided into two categories.

I.1 Crystalline solid

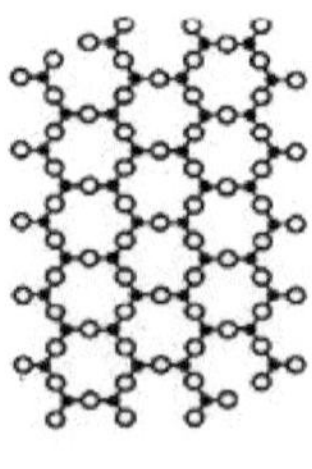

Figure 1: SiO_2 (**crystalline**) [1].

It corresponds to a regular assembly (***ordered state***) on a microscopic scale (Fig.1), i.e. a repetition of a basic unit over large distances compared with the interatomic distances. They are characterised by :

A purely defined crystallisation and melting temperature. (Change of state stage under fixed pressure). Anisotropic physical properties linked to their structure (e.g. electrical conduction or ease of cleavage in one specific direction).
✓Resistance, elasticity, electrical conductivity, thermal conductivity...etc.

1.2.Amorphous solid

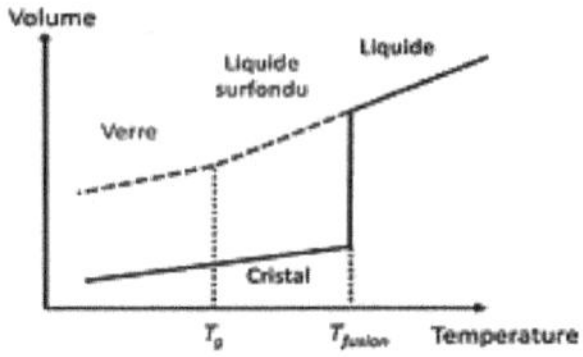

2. Figure 2: Variation in volume as a function of temperature for amorphous and crystalline solids [2].
3. Glass transition temperature: Tg=1473 K.
4. Melting temperature: Tmelt=1996 K.

It is not spatially organised (disordered ***state)***. ***disordered state***) (Fig.3) of the entities
(ions, molecules).

✓ There is no critical temperature melting point
✓ It gradually softens on heating and turns into a viscous liquid.
✓ The physical properties of an amorphous body are isotropic (the same in all directions).

Example: glass, polymers, etc.

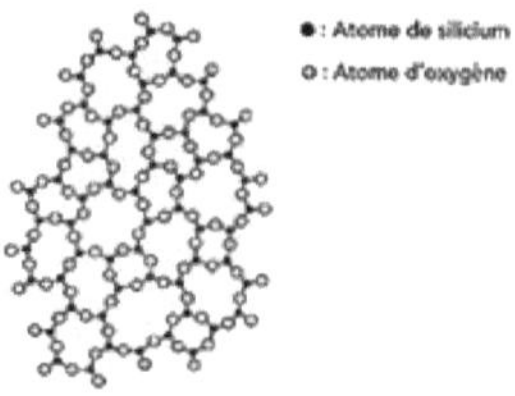

Figure 3: SiO_2 in glass (**amorphous**). [1]

4.2.Perfect crystal

In crystalline solids, the organisation of the atoms is perfectly regular. The purely theoretical state of the **perfect crystal** is that of a crystal repeating itself infinitely identically in all directions in space, which also implies the absence of any impurity (***state of perfect order***). All solids are actually situated between these two limit states (perfectly regular and purely random).

4.3.Classification of the various types of crystals

Three main factors, which are not independent, determine the type of crystal:

- ***The chemical factor***

This is linked to the positions in the periodic table and therefore to the relative electronegativities of the atoms in the crystal, which are decisive for the valence bond.

- ***The steric factor***

Due to the volume of atoms and ions; whose influence is predominant in the formation of ionic crystals.

- ***The electronic factor***

This is due to the number of electrons in the valence layers, which has the greatest influence in covalent structures.

Crystals are classified according to the interaction responsible for crystal cohesion:

a) Strong bond crystals

- **Metal crystals**

The metal bond (Fig.4) is established between atoms with low electronegativities and few electrons on their outer layer (1, 2 or 3 electrons).

- pooling of electrons not on 2 atoms but on an unlimited number of atoms: phenomenon of delocalisation of electrons throughout the sample - not directed in space.

Model of the metal bond :

When the bond is formed, the metal atoms lose their influence over their outer electrons: they therefore become positive ions whose positions, if the metal is solid, are fixed in relation to each other. The external electrons are delocalised and behave as if they were free, while remaining in the sample.

A metal can be described as an assembly of positive ions bathed in a weak electron cloud (or sea) whose electrons are easily mobile, hence the high electrical conductivity of metals.

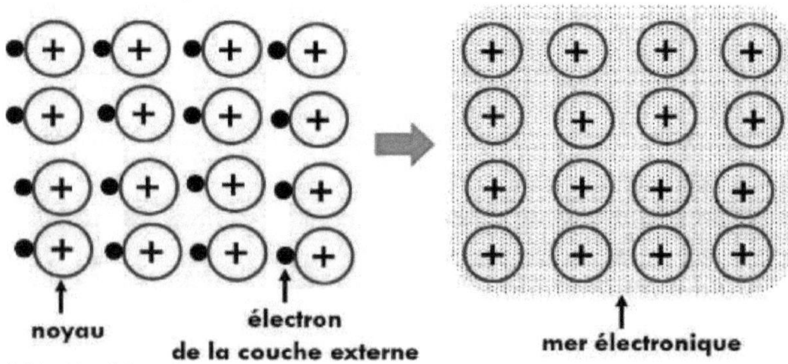

Figure 4: Metal bond formation

Example: Fe(s), Cu(s) and metal alloys such as Ag_xCu_y.

- **Ionic crystals**

Ionic bonding (Fig.5) results from electrostatic interactions between ions of opposite charge.

When the difference in electronegativity between the two atoms is large enough, the first atom is able to capture valence electrons from the other atom. The result is a positive ion (low ionisation energy) and a negative ion (high electron affinity). Since opposite charges attract, the two ions are bound to each other. An atom that gives up valence electrons has only a small number of electrons on its outermost layer, and often eight electrons on the penultimate layer. When it gives up its valence electrons, it reaches the rare gas configuration.

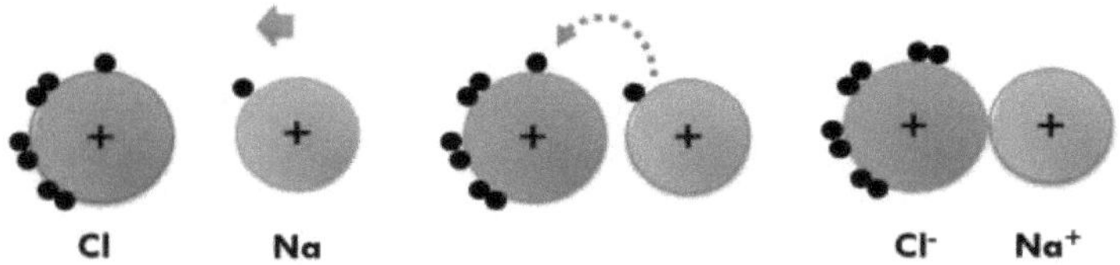

Figure 5: Ionic bond formation in the NaCl(s) crystal.

Example: CsBr(s), NaCl(s), CaF_2, CaO, etc.

- **Covalent crystals**

The covalent bond (Fig.6) is obtained by **the overlapping of atomic orbitals** associated with the valence electrons of the linked atoms. The overlap gives rise to collective electron states in which the potential and kinetic energies of the electrons are minimised.

The density of the probability of electrons being present in the bonding states is maximum between the atoms bonded together. Thus, the covalent bond can be seen **as the sharing of two electrons by the linked atoms, with each atom contributing one electron.**

Covalent bonds **are directional**. The geometry of the bonds is determined by the angular dependence of the atomic orbitals, whether pure or hybrid [3].

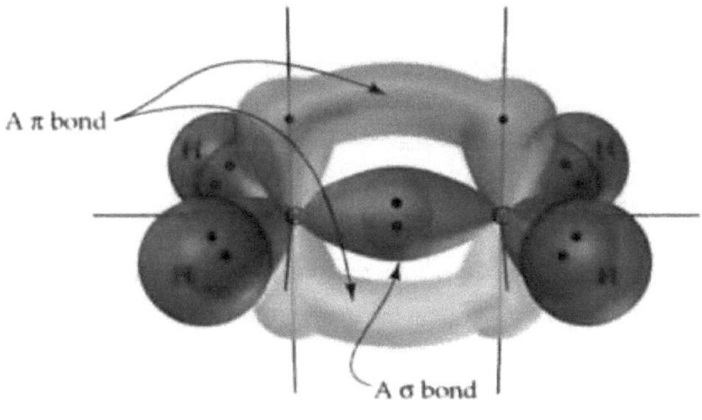

Figure 6: Covalent bond formation.

Ion-covalent bonds

A large number of compounds have bonds with both ionic and covalent characteristics. A covalent bond involving two atoms of different chemical natures cannot be perfectly symmetrical, mainly because of the difference in electronegativities. In this case, the electronic distribution of the bond is shifted towards the atom with the stronger electronegativity, giving the bond an ionic character.

Example: C(s), SiO2 (quartz).

b) Weakly bonded crystals

- **Molecular crystals**

Hydrogen bond

- occurs when an electronegative atom (with one or more free doublets) is in the vicinity of a hydrogen atom covalently bonded to another electronegative atom.
- directed into space.
- intramolecular and intermolecular H-bonds

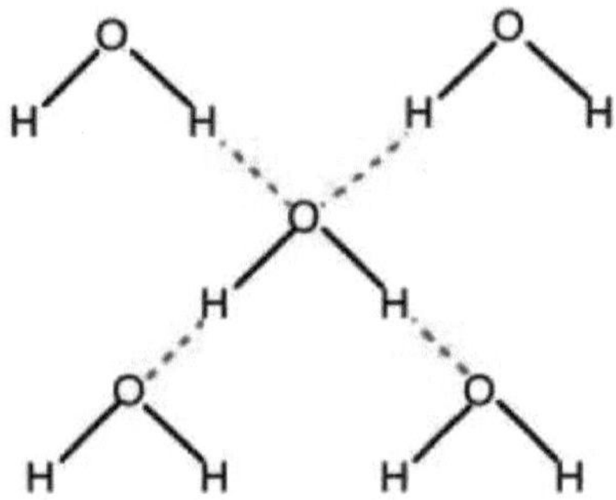

Hydrogen bond formation. [4]

[4] K. Ilhem, chemical bonding, Farhat Abbas University.

Van der Waals bonds

- generally very low
- arise from the attraction between permanent electric dipoles (for polar molecules) or induced dipoles in atoms or molecules.
- not directed in space.

3 types of Van der Waals bonds :

- Attraction between permanent dipoles in polar molecules: ***Keesom forces***
- Attraction between permanent dipoles (polar molecules) and dipoles induced in non-polar molecules (induced by permanent dipoles in polar molecules): ***Debye forces***.
- Attraction between non-polar molecules, due to the polarizability of

molecules or atoms: the most fundamental interaction of the 3, since it still exists: ***London dispersion forces.***

Example: I_2(S), CO_2(s) and H_2O(s).

1.4.Size

In most metallic, ionic and covalent crystalline solids, the strong bond is **three-dimensional (3D)**. This is the case for silver, sodium chloride and diamond. This gives the materials certain properties (high melting point, hardness, etc.). Some solids have a **two-dimensional (2D)** configuration involving parallel planes called "sheets". The interatomic bonds in the sheets are stronger than those between them. These solids are easily cleaved (graphite variety of carbon). Some crystallised materials have **a one-dimensional (1D)** structure, with strong bonds appearing along a single axis. This is the case with DNA polymer chains.

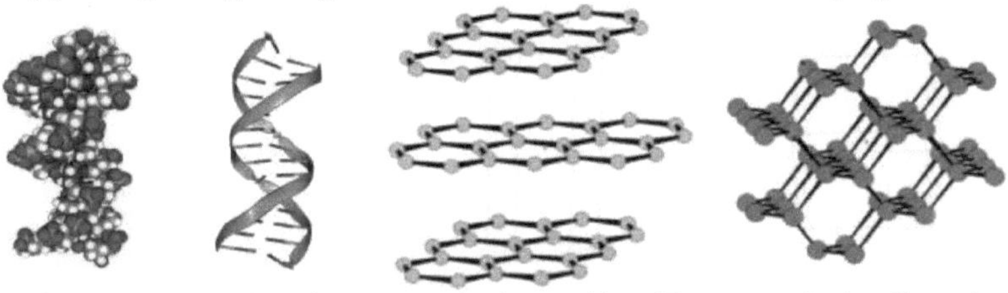

Structure of DNA: **One-dimensional (1D)** 2D structure of carbon graphite: **Two-dimensional (2D)** 3D structure of carbon diamond : **Three-dimensional (3D)**

II. Basics of crystallography

1.1.Description of a crystal

a) Design

The pattern is the smallest discernible entity that is repeated periodically by translations in the crystal (atom, ion, molecule, etc.).

b) Reseau

The description of the crystal structure can be simplified by replacing the patterns with points called mends. The periodic arrangement of mends is called the crystal lattice.

The origin of the network can be chosen arbitrarily.

The repetition period of the pattern is defined by two or three non-collinear vectors *a, b* and *c* , depending on whether the network is two-dimensional or three-dimensional.

Any transition of vector $t = ua + vb + wc$ ***with u, v and w relative integers, brings the periodic system into coincidence with itself.***

Crystal structure = crystal lattice + pattern.

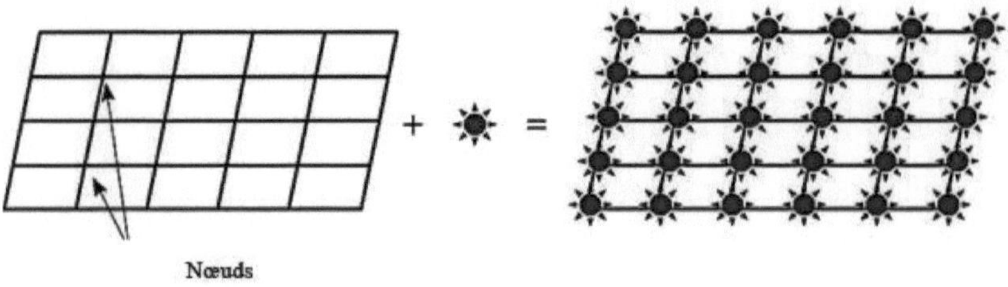

Nodes
Crystal lattice Pattern Crystal structure

c) Mesh

A mesh is a finite part of space, by translation of which the crystalline pattern can be reproduced ad infinitum.

In a three-dimensional crystal structure, the mesh is the smallest parallelepiped constructed from an origin and the 3 translation vectors *a, b* and *c* (direct reference). It is defined by 6 parameters (3 lengths and 3 angles in the crystal.

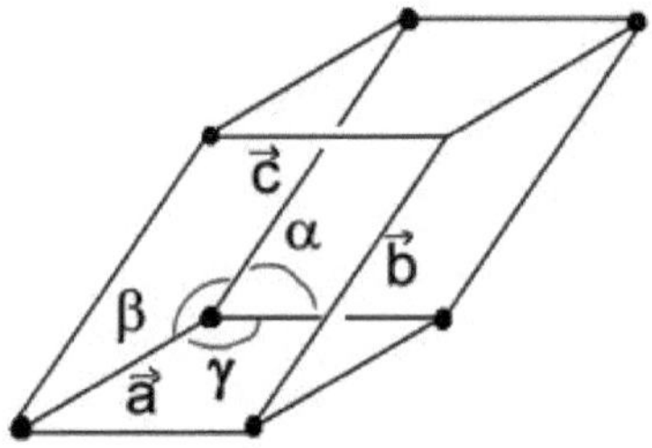

Mesh parameters: 3 vectors $(\vec{a}, \vec{b}, \vec{c})$ and 3 angles (α, β, γ).

The translations are chosen so as to move from the pattern to an identical pattern.) This pattern can be an atom, a molecule or a group of atoms, molecules or ions. In most cases, it corresponds to the chemical formula (unit or form group).

A stitch perhaps:

✓ **Simple**: the nodes of the network are merged with the mesh vertices.

✓ **Multiple**: the mesh also has naps at other points.

The volume of the mesh can alternatively be defined as the scalar product of a base vector and the vector product of the other two vectors:

$$V = (\vec{a} \wedge \vec{b}).\vec{c} = (\vec{a} \wedge \vec{c}).\vec{b} = (\vec{b} \wedge \vec{c}).\vec{a}$$

Since the volume of a mesh is positive, the base vectors must form a direct triedre.

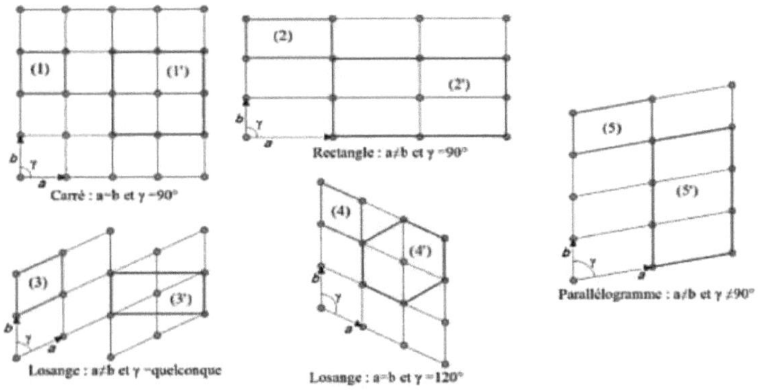

The different types of mesh in a two-dimensional network.

11.2. Crystalline systems and network types

Each type of elementary mesh (characterised by certain relationships between the mesh parameters) is associated with a **crystalline system.** There are 7 different **crystalline systems**.

In addition, we can define 4 types of network, characterised by the way in which the nodes are distributed in the mesh:

✓ **Primitive (P)**: A simple mesh is called **primitive** and is denoted **P.** It contains 8 net ids at the vertices, each of which accounts for 1/8 of the mesh, i.e. 8 x 1/8 = **1** node per mesh. The only network translations are **linear** combinations **of the 3 base vectors** *(a, ~b,* c).

✓ **Centre (I):** A mesh containing a node at the vertices of the mesh and a node at the centre of the mesh, with a total of 2 nodes per mesh, is known as a **centred** mesh, denoted **I**. For this type of mesh, the translation **(% % %) must be** added to the **linear** combinations **of the 3 basic vectors** (cl, *b, c).*

✓ **All centred faces (F)**: A mesh containing one node at the vertices of the mesh and 3 nodes on the 6 centred faces (6x1/2= 3), giving a total of 4 nodes per mesh, is said to have centred faces, denoted **F**. For this type of network, we need to add the translations **(% % 0), (% 0 %)** and **(0 % %)** to the **linear** combinations **of the 3 base vectors** *(a,* b, c).

✓ **2 opposite faces centred C** (**A** or **B**)**:** A mesh containing a node at the vertices of the mesh and a node on 2 centred faces (2x1/2 = 1), a total of 2 nodes per mesh. This is known as a 2-sided mesh and is labelled **C** (**A** or **B**). For this type of mesh, we need to add the translations **(% % 0), (% 0 %)** and **(0 % %)** to the **linear** combinations **of the 3 base vectors** *(a*, b, c). For mode C, add the translation **(% % 0)** to the **linear** combinations **of the 3 base vectors** (1, b, c).

✓ If we combine the 7 crystalline systems with the 4 possible lattice types (not all lattice modes are compatible with the symmetry of the systems), we define **14 lattices** known as **Bravais lattices**. For the **cubic** system, there are only the 3 modes P, I and F. For the **hexagonal** system, there is only mode P.

Table 1. The 7 crystal systems and 14 Bravais lattices.

Crystalline system	**Mesh parameters**	**Possible network modes (14)** **P I F C**
Triclinic	a≠b≠c α ≠ β ¯© ≠ 90° Any parallelepiped	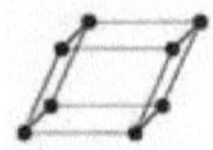
Monoclinic	a≠b≠c α = γ = 90°, β ≠90° Right prism with parallelogram base	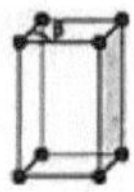 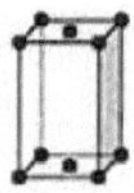
Orthorhombic	a≠b≠c α = β = © = 90° Trirectangular parallelepiped	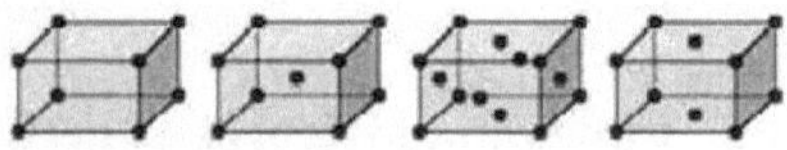
Quadratic	a=b≠c α = β = © = 90° Straight prism with square base	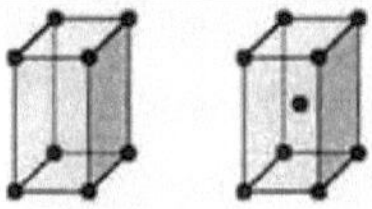
Cubic	a=b=c α = β = © = 90° Cube	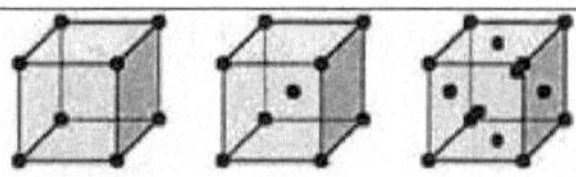
Rhombohedral	a=b=c α = β = © ≠ 90° and<120° rhombohedron	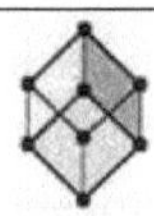
Hexagonal	a=b≠c α = β = 90° ; γ = 120° Right prism with rhombus base	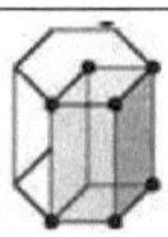

For the purposes of this course, we will confine ourselves to the cubic and hexagonal systems.

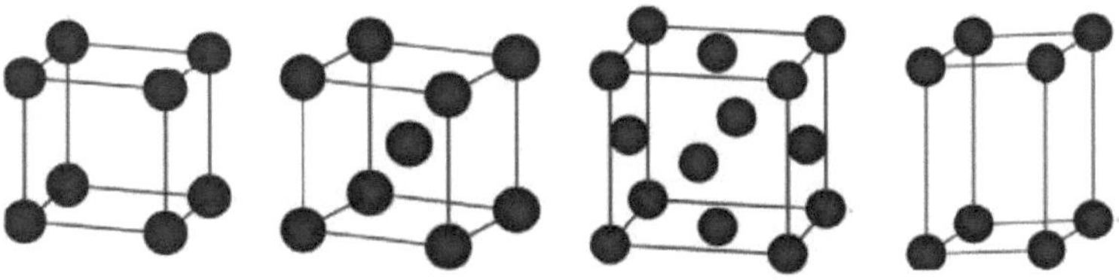

Cubique simple (P) | Cubique centré (I) | Cubique à Faces Centrées (F) | Hexagonal simple (P)

Simple cubic
(P)
Centred cubic
(I)
Cubic with Centred Sides
(F)
Single hexagonal
(P)

II.3 Elements of symmetry in finite figures

a) Inversion

Note 1, relative to a point, the centre of inversion, i any point of x, y, z coordinates has a symmetrical point of coordinates -x, -y, -z the mesh is then said to be centro-symmetrical.

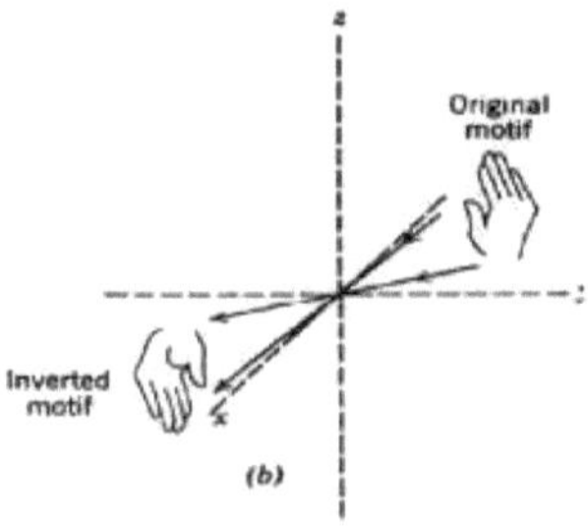

b) Reflection

Image in a mirror or plane of symmetry **m**.

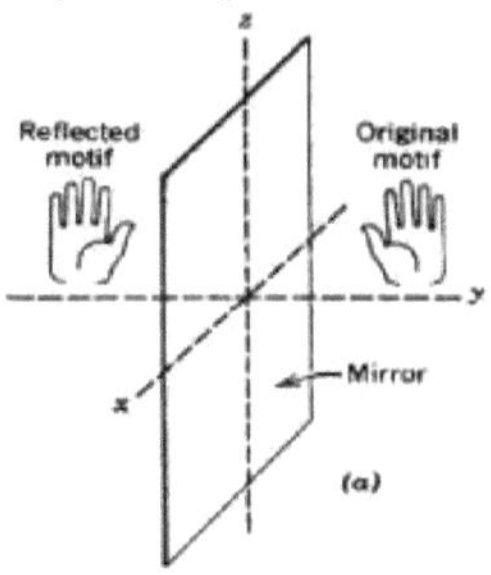

c) Rotation of order n

Rotation of 2n/n, around an axis of rotation represented by the value of n. Possible rotations are for n=1 (identity), 2, 3, 4 and 6. Possible rotations are for n=1 (identity), 2, 3, 4 and 6. (Axes n=5 and n>6 are impossible in crystals.

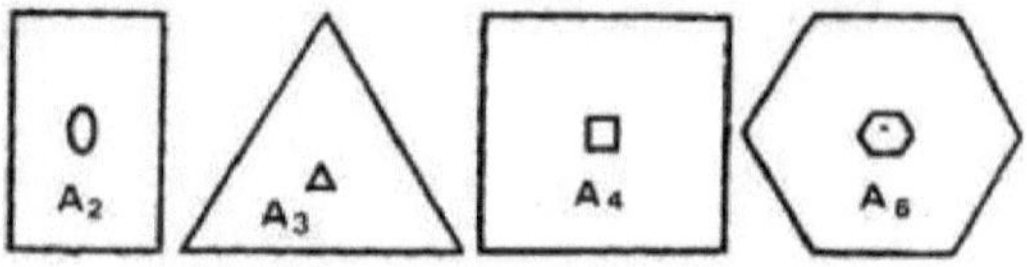

Example : Elements of symmetry in the cube

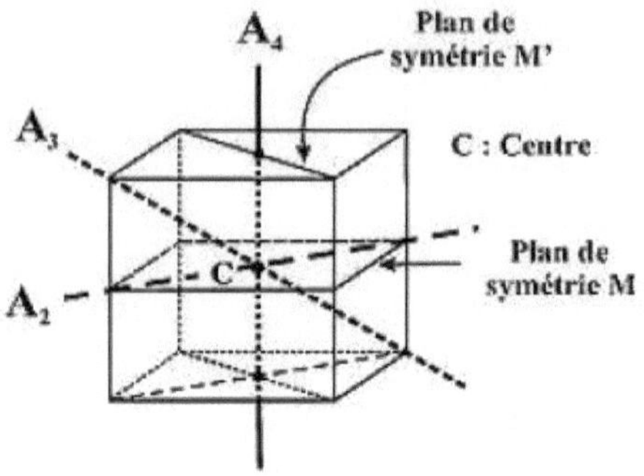

3A₄/3M ; 4A₃ ; 6A₂/6M

II.4. Ranges

A range is a straight line which passes through the origin and the coordinate node (uvw). A straight line parallel to the range [uvw] passes through each node in the grid. The set of parallel and equidistant lines forms the [uvw] range family.

By convention u, v and w must be mutually prime.

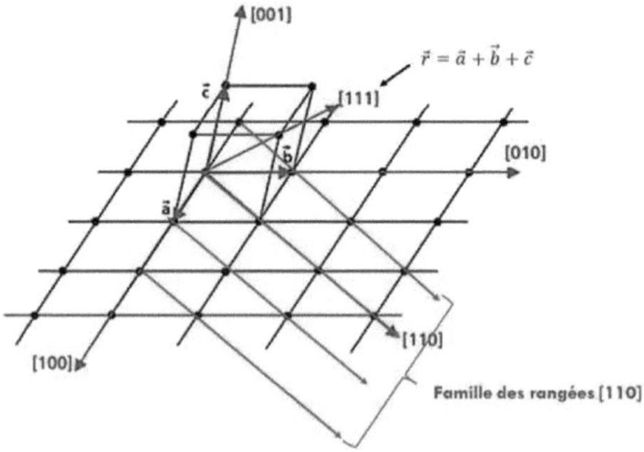

II.5. Reticular planes

A **reticular plane** is defined by three non-aligned nodes; it contains other iHL'uds whose two-dimensional distribution is periodic. It is designated by ***Miller indices (h k l)*** which are positive, negative or zero integers. They are such that the corresponding plane cuts the edges ***a*** into ***a/h***, ***b*** into ***b/k*** and ***c*** into ***c/l***.

The Miller indices (hkl) are mutually prime integers and represent a family of parallel reticular planes. The first plane in the family is defined by the indices (hkl).

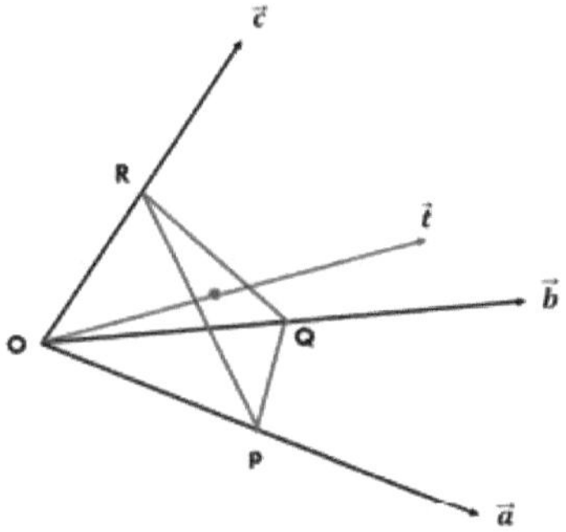

OP=a/h
OQ=b/k
OR=c/l
h, k and l are mutually prime relative integers.
These are called Miller indices.
The PQR reticular plane has indices (222).

- Family of planes (hkl) verifying the equation: hx+ky+lz=m with m integer.
- m=0 the plane passes through the origin O
- m=1 or -1 : 1ere planes // and closest to the origin.

The plane with m=1, P_1, intersects the directions [100], [010] and [001] in fractions of the base vectors of the network: a/h, b/k, c/l.

For m=n, plane P_n cuts na/h, nb/k, nc/l.

If the plane P1 is // a base vector of the network, the corresponding index will be 0 ; P1// $\vec{a}$. then h=0.

Application

Represent the following reticular planes:

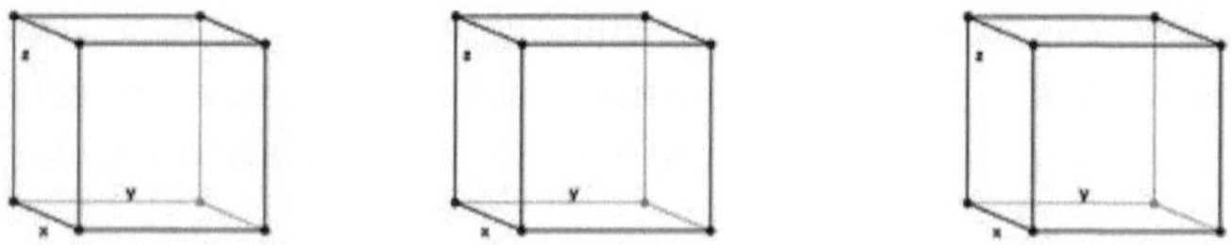

(100) (110) (310)

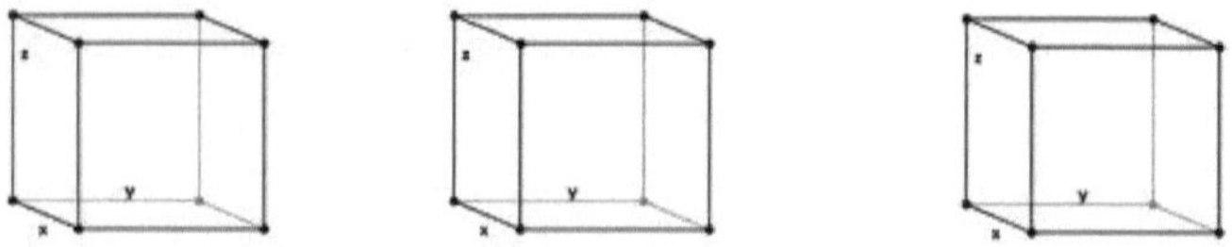

$(2\bar{1}\bar{1})$ $(\bar{1}10)$ (222)

The distance separating two successive planes in the same family of reticular planes is called the inter-reticular distance. It is noted: dhkl.

Simple cubic network: $d_{hkl} = \frac{a}{\sqrt{h^2+k^2+l^2}}$

Examples of dhkl (Case of the cubic system)

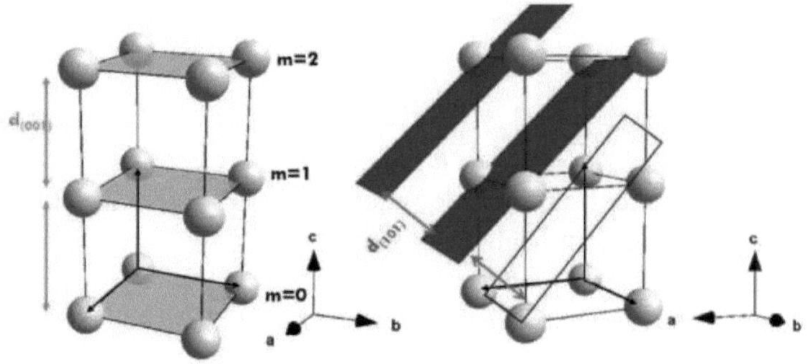

Family of reticular planes (001) and (101).

III. X-rays and the diffraction phenomenon

III.1. 1. X-rays

X-rays are the basis of various analysis techniques such as radiography, spectroscopy and diffractometry. This electromagnetic radiation has a wavelength of the order of one Angstrom (1 A = 10^{-10} m).

A crystal is an arrangement of atoms, ions or molecules with a periodically repeating pattern in three dimensions. The interatomic distances are of the order of Angstrom, the same order of magnitude as the wavelengths of X-rays: a crystal therefore constitutes a 3D lattice that can diffract X-rays.

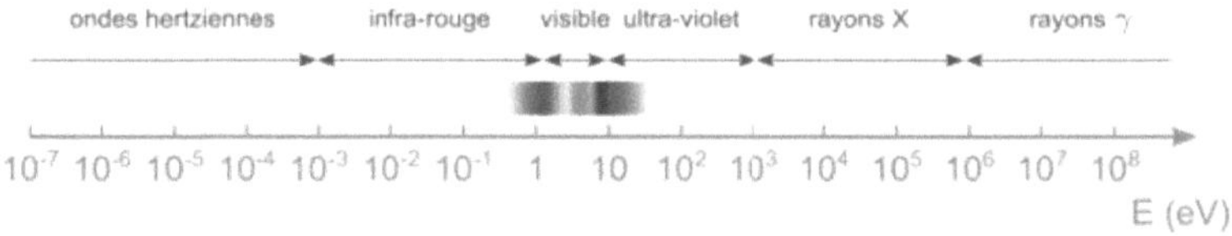

We present the basic theory of the interaction of X-rays with solid structures, as well as examples of applications: the resolution of crystalline structures on powder.

III.2. 2. X-ray diffraction: Bragg's law [4].

The general method consists of bombarding the polycrystalline sample with X-rays, and observing the intensity of the X-rays, which is scattered according to the orientation in space of all the atoms in the families of reticular planes. The scattered X-rays interfere with each other, so the intensity has *maxima* in certain directions; this is known as the "diffraction" phenomenon. The intensity detected is recorded as a function of the angle of deviation 2θ ("two-theta") of the beam; the curve obtained is called the "diffractogram".

Example:

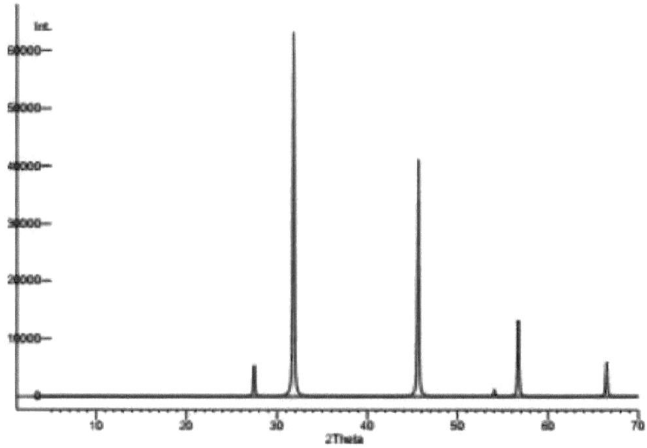

NaCl diffractogram.

Diffraction can be thought of as reflection from parallel reticular planes.

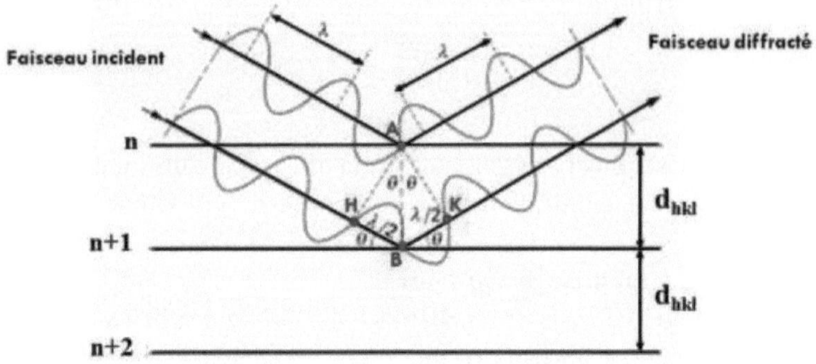

This implies that : $BH = BK = d_{hkl} \sin\theta$

Constructive interference occurs if and only if the two rays 1 and 2 are in phase, i.e. : $BH + BK = n\lambda$

Hence : $2d_{hkl} \sin\theta = n\lambda$ **Bragg's law**

n is an integer which defines the order of the diffraction. In practice, we limit ourselves to order 1.

IV. Application of X-ray diffraction (Debye-Scherrer method)

This is the most commonly used method when the material can be reduced to a fine powder (grains of the order of 0.01mm). The sample is a point sample and the diffracted monochromatic X-rays are recorded on a circular film centred on the sample.

IV.1 Principle of the method

The X-ray beam, which is **monochromatic** in this case, falls on the microcrystalline powder on a small glass rod, in a small capillary (Lindemann capillary) or spread on a special thin slide.

The basic assumption is that among all the small crystals (microcrystals) present (in principle non-oriented) there will be enough to present crystalline faces such that diffractions can be made at Bragg angle 2θ. We will record

Introduction and basic notions of crystallography done the diffracted rays on a photographic film. The direct beam (emergent ray in the extension of the incident ray) is absorbed in another tube called a well.

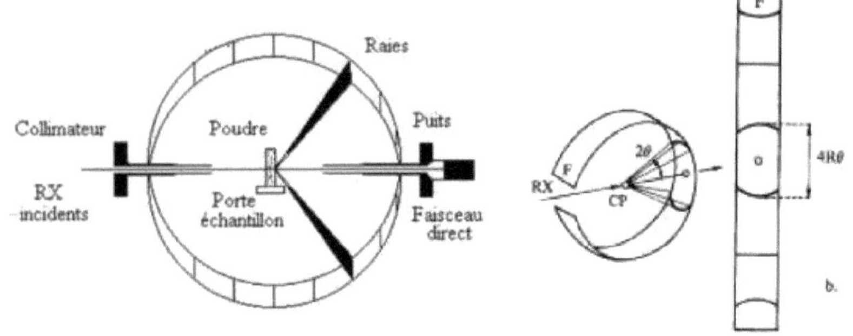

Figure 1: Principle of the Debye-Scherrer method.

With a circular Debye-Scherrer chamber, concentric rings are obtained on the film, each representing a reticular distance (the diffracted rays produce a series of cones and intersect the cylindrical film. Curves are thus obtained which are the intersection of these cones with the cylinder (these curves are called rings). The holes in the film come from the collimator and/or the well, depending on the set-up. Note the symmetrical arrangement of the lines in relation to both the collimator and the well.

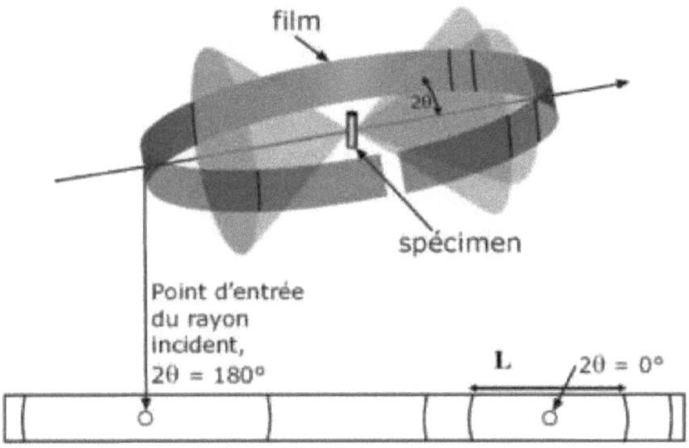

Figure 3: Appearance of the film. [5].

IV.2 Interest of the method

This diagram enables :

- Calculate the equidistances dhki of the different planes: 40R= L with 0: diffraction angle in (°) and L: distance between two symmetrical lines in mm. Chamber circumference $C=2\pi R$ ((C= 360 mm or 240 mm or 180 mm).

Calculate the diffraction angle on 0i corresponding to each pair of lines.

d) C=360 mm donc $\theta_i = \frac{L_i}{4}$

e) C= 240 mm donc $\theta_i = \frac{3L_i}{8}$

f) C= 180 mm donc $\theta_i = \frac{L_i}{2}$

By measuring the distance between two symmetrical lines, we can calculate the inter-reticular distance, using Bragg's relation ; $2d\sin\theta = n\lambda$.

➢ $d_i = \frac{\lambda}{2\sin\theta}$ (Å)

- Determine the type of grating and the parameters of the mesh, by indexing the lines, i.e. determine to which plane *h, k, l* each line belongs.

$$d_{hkl} = \frac{a}{\sqrt{h^2+k^2+l^2}}$$

IV.3 Line indexing of a powder diffractogram: cubic gratings

The main application of this powder method is the identification of solid substances. Identification is made possible by means of data sheets. Each sheet contains at least the reticular distances and the corresponding relative intensities. They may also contain the Miller indices (h k l), the d_{hkl}, the crystalline system and the lattice mode. Analysis of the film consists of comparing the data obtained with those in the data sheet for the substance in question.

Reporting methods

This method consists of determining the lattice mode of the crystalline system of the polycrystalline sample under examination.

The indices of the diffracting planes for a compound crystallising in the cubic system are in ascending order of h +k +l^{222} :

- **Example of a cubic system:**

The lists of reflections and corresponding distances for the three types of cubic networks are :

****In the case of a simple cubic lattice P,*** lines are observed in all the reticular planes: (100); (110); (111); (200); (210) ...etc.

$d_i : a \; ; \; {}^{a}/_{\sqrt{2}} \; ; \; {}^{a}/_{\sqrt{3}} \; ; \; {}^{a}/_{\sqrt{4}} \; ; \; {}^{a}/_{\sqrt{5}}$

Case of a cubic lattice centre I: lines are observed for the reticular planes with indices verifying the sum **h+k+l=2n** (110); (200); (211); (220); (310); (222), etc.

$d_i : {}^{a}/_{\sqrt{2}} \; ; \; {}^{a}/_{\sqrt{4}} \; ; \; {}^{a}/_{\sqrt{6}} \; ; \; {}^{a}/_{\sqrt{10}} \; ; \; {}^{a}/_{\sqrt{12}}$

In the case of a face-centred cubic lattice F: lines are observed for the reticular planes with indices verifying the condition **h, k and l of the same parity**.

(111) (200) ; (220) ; (311) ; (222) .etc

$d_i : {}^{a}/_{\sqrt{3}} \; ; \; {}^{a}/_{\sqrt{4}} \; ; \; {}^{a}/_{\sqrt{8}} \; ; \; {}^{a}/_{\sqrt{11}} \; ; \; {}^{a}/_{\sqrt{12}}$.

Table of ratios ($_{d1/dhkl}$) with increasing values of (h +k +l)^{222}

H	k	L	(h +k +l)^{222}	Cubic P ($_{d1/dhkl}$)^2	Cubic I 2($_{d1/dhkl}$)^2	Cubic F 3($_{d1/dhkl}$)^2
1	0	0	1	1	-	-
1	1	0	2	2	2	-
1	1	1	3	3	-	3
2	0	0	4	4	4	4
2	1	0	5	5	-	-
2	1	1	6	6	6	-
2	2	0	8	8	8	8
2	2	1	9	9	-	-
3	0	0	9	9	-	-
3	1	0	10	10	10	-
3	1	1	11	11	-	11
2	2	2	12	12	12	12
3	2	0	13	13	-	-
3	2	1	14	14	14	-
4	0	0	16	16	16	16
3	2	2	17	17	-	-
4	1	0	17	17	-	-
3	3	0	18	18	18	-
4	1	1	18	18	18	-
3	3	1	19	19	-	19
4	2	0	20	20	20	20

The numbers 7 and 15 are always absent because they do not correspond to any sum of squares of integers.

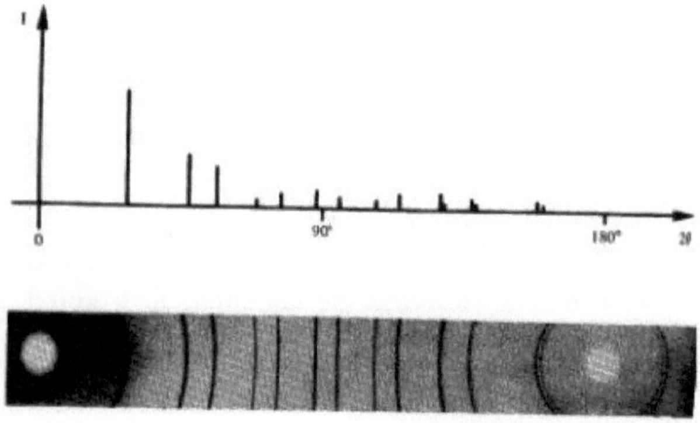

Comparison between a Debye Scherrer plate and a powder diffractogram.

References :

[1] C. Salem, 2010, Elaboration et caracterisation de nouveaux verres de fluorohafnates de strontium et de phosphosulfates, thesis, Universite de Mohamed Khider Biskra.

[2] **https://physique-et-maths.fr/soutien-scolaire.php?menu=416.**

[3] C. Laulhe, structure of matter, Université Paris Saclay.

[4] OUDET, X. (2013). Bragg's law, the continuous and the discontinuous. In Annales de la Fondation Louis de Broglie (Vol. 38, No. 1, p. 33).

[5] M.Corentin, G. Adrian, L3PAPP practical crystallography work.

CHAPTER 2

Metal crystals

CHAPTER 2

Metal crystals

I. General

The simplest model is to represent the metal crystal as an ordered stack of fixed metal ions between which an electron cloud can circulate, delocalized to the whole crystal. Metals and metal alloys tend to give high density structures.

For example: Ba, y-Fe, W: face-centred cubic structure (I); Ni, Cu, Ag, Au, a-Fe, Al: face-centred cubic structure (CFC): Be, Mg, Zn, Cd: compact hexagonal structure (HC).

For the metal iron: example of the existence of ***allotropic varieties***, ***different structures each with its own range of stability for the same chemical compound.***

A compact metallic structure is described as a stack of rigid spheres (impenetrable and indeformable (atoms assimilated to spheres) ensuring maximum filling of space. Two types of maximum packing plane stack can be constructed.

ABCABCABC stacking...

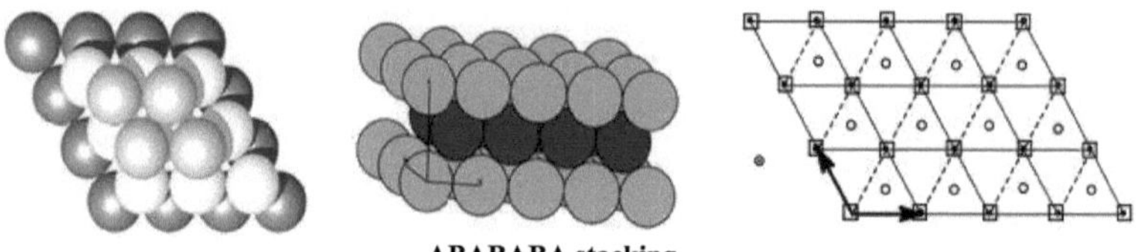

ABABABA stacking...

In each type of stack, each sphere is in contact with 12 other spheres: 6 from the same plane, 3 from the previous plane and 3 from the next plane. The co-ordination (or co-ordination index) is said to be 12. To find the nodes, draw the centres of these spheres (without forgetting that there is matter, and therefore without forgetting the existence of directions in which the spheres are tangent). In reference to the nature of the mesh that appears, these ***two types of stack*** are called respectively: ***face-centred cubic structure (sequence ABC-ABC-ABC)*** and ***compact hexagonal structure (sequence AB-AB-AB).***

II. Face-centred cubic structure

The face-centred cubic (CFC) structure is the result of an infinite stack of A, B and C planes.

II.1. . Mesh

The mesh of a face-centred cube is of multiplicity 4. The mesh contains 4 atoms divided into 8 atoms at the vertices (the sphere centred on such a node is divided between 8 meshes) and 6 at the centres of the faces (the sphere centred on such a node is divided between 2 meshes). So in total ***(8 *1/8+ 6*1/2)=4.***

The reduced coordinates of the atoms are: (000); (1/2 1/2 0); (0 1/2 1/2); (1/2 0 1/2).

The planes A, B and C of the stack are arranged perpendicular to the long diagonals of the cube (rangee [111]: Axis of rotation of order 3).

Compact model

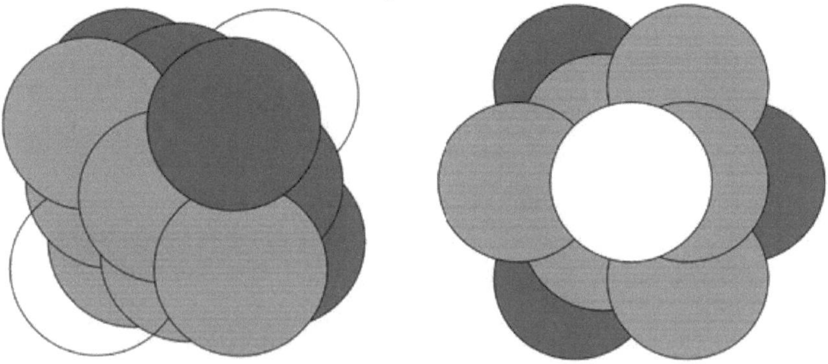

Perspective representation Projection in the (111) plane

Exploded model

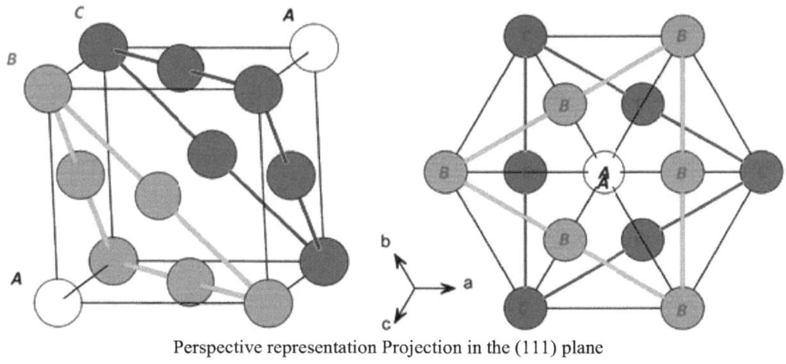

Perspective representation Projection in the (111) plane

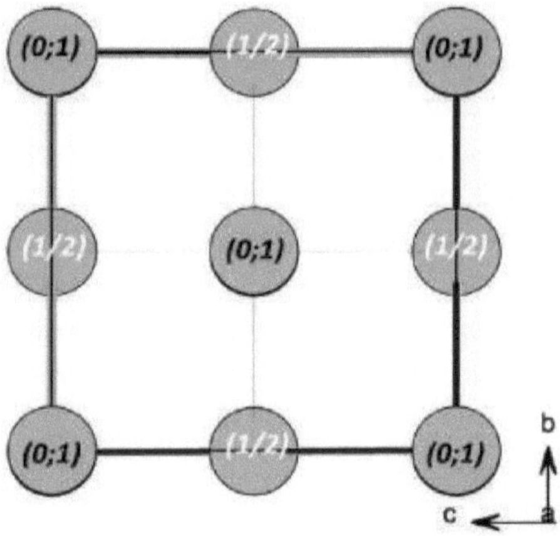

Projection of a CFC mesh in the (100) plane.

II.2. . Contact

The co-ordination of an atom or ion is the number of nearest equidistant neighbours. Each sphere is in contact with 12 other spheres: 6 in the same plane, 3 in the previous plane and 3 in the next plane. The co-ordination (or co-ordination index) is 12.

Note: the coordination index **IC** describes the coordination polyhedron of the central atom. The most common values of **IC** are: 4, 6, 8, (Ic = 4: tetrahedron, Ic = 6: octahedron, Ic = 8: cube). To draw this polyhedron, we need to trace the edges connecting 2 close neighbours.

II.3. . Direction of sphere tangency and relationship between mesh dimensions and atom size

Beware of the difference in the way compact models are used (visualising the filling of the space occupied by matter) and exploded models (visualising the locations where the centres of the spheres are located). Relationship between metal radius **r** and mesh edge **a**: find the directions in which the spheres are tangent (at contact).

Table 2. Relationship between the cubic mesh parameter a and the particle radius r

Structure	Next tangent	Relationship
Single cubic (CS)	A arete	a = 2r
Centred cubic (CC)	The diagonal of the cube	4r = a^3
Face-centred cubic (CFC)	The diagonal of a cube face	4r = a^2

For a face-centred cubic mesh we have : $4r = a\sqrt{2} \quad a = 2r\sqrt{2}$

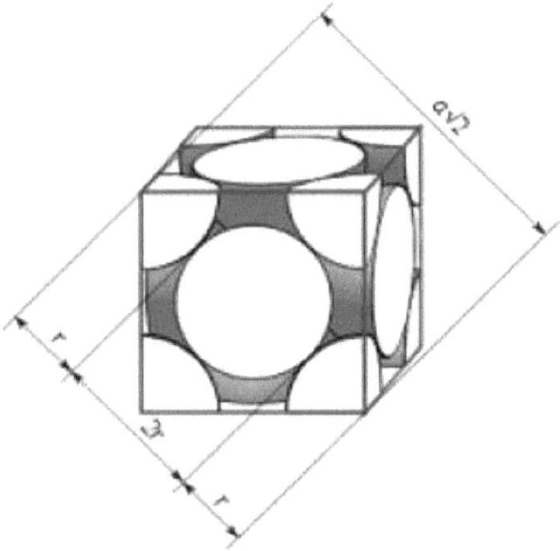

Atomic tangents in a compact CFC mesh. [1]

II.4. . Compactness

The compactness C of a solid represents the ratio of the volume Vp actually occupied by the particles of matter (atoms, molecules, ions) to the volume of the crystal, the overall volume actually occupied in space. It is generally calculated for a crystal volume equal to that of an elementary mesh Vm. Under these conditions, C = Vp / Vm.

Compactness of a metal structure with maximum compactness: C = 0.74=74%.

$Compacite = \frac{Z \times V_{sphère}}{V_{maille}} = \frac{4 \times \frac{4\pi r^3}{3}}{a^3}$; On a : $4r = a\sqrt{2}$ $\quad a = 2r\sqrt{2}$ therefore

$Compacite = \frac{\pi \times \sqrt{2}}{6}$. ***. Mass density***

The density **p of** a material corresponds to the mass of an element per unit of volume. It should therefore be expressed in $kg.m^{-3}$ in the international system, but we prefer to use the units $g.cm^{-3}$.

The volume of the mesh is: $V_{atomes} = a^3$

It comprises 4 atoms (Z=4) whose mass is : $m_{atomes} = \frac{4 \times M}{N_A}$

The density is expressed as follows: $\rho = \frac{m_{atomes}}{V_{atomes}} = \frac{4 \times M}{N_A \times a^3}$

Conversely : $a = \sqrt[3]{\frac{4 \times M}{N_A \times \rho}}$

Application: Calculate the parameter of a cubic copper lattice of lattice mode F, given that p = 8.96 g/cm3 . M(Cu) = 63.54 $g.mol^{-1}$.

II.5. . Crystallographic sites

The C.F.C structure reveals two interstitial cavities, or crystallographic sites, which are of two types: ***octahedral sites (O_h)*** located <u>at the centre of a regular octahedron whose vertices are the centres of 6 spheres tangent to each other and at equal distance ***a/2*** from the centre.</u> ***Tetrahedral sites (T_d)*** located <u>at the centre of a regular tetrahedron whose vertices are the centres of 4 spheres tangent to each other and at equal distance</u> $\mathbf{a\sqrt{3}/4}$ <u>from the centre</u>. The size of the site is measured by the maximum radius of the atom that would occupy the site without deforming it.

i. Octahedral sites	ii. **Tetrahedral sites**
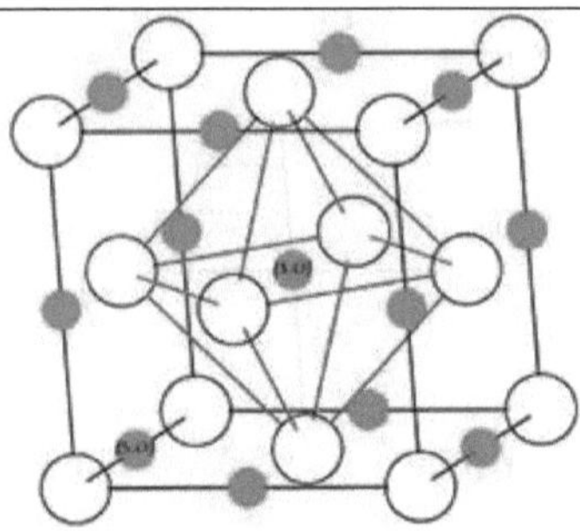 The octahedral sites (***S.O***) of a CFC mesh.	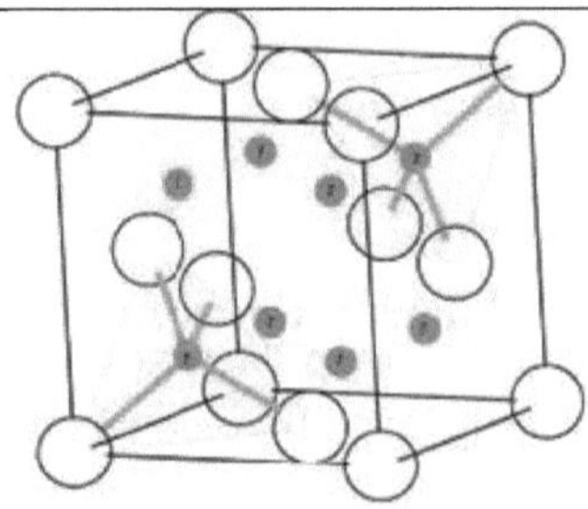 Tetrahedral sites (***T***) in a CFC mesh.
There are 4 octahedral sites per C.F.C mesh. They are located in the centre of the mesh and in the middle of the edges. The octahedral site in the centre belongs entirely to the mesh and those on the edges (12 in all) are each shared between 4 neighbouring meshes; hence 1+12×1/4= 4 octahedral sites per mesh. ***<u>Their positions have the following coordinates</u>:*** **(1/2 1/2 1/2); (1/2 0 0); (0 1/2 0); (0 0 1/2).** In the C.F.C. structure, there are as many octahedral sites as atoms per cell. The size of the octahedral site is given by the	In the C.F.C structure, there are 8 tetrahedral sites per cell. They are positioned at the centre of regular tetrahedra defined by one atom at the vertex of the mesh (atom of plane A) and 3 atoms at the centres of faces passing through this vertex (atoms of plane B). These positions correspond to the centres of the eight cubes with **a/2** edges. All these 8 sites belong completely to the mesh. The number of tetrahedral sites is twice the number of atoms per mesh. ***<u>Their positions have the following</u>***

following relationship: 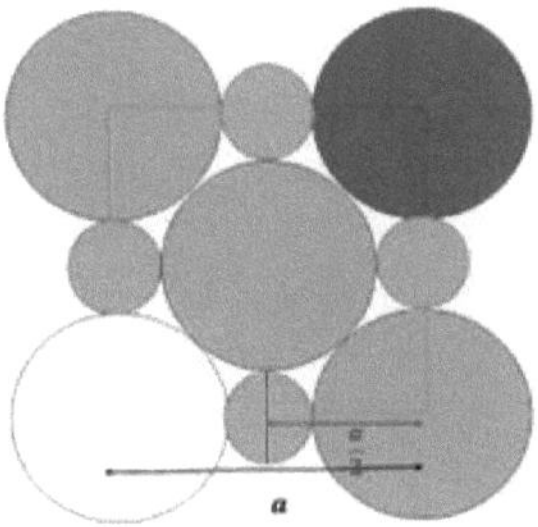 $$2r + 2R_O = a \ avec \ 4r = a\sqrt{2} ;$$ $$R_O = \frac{2r}{\sqrt{2}} - r = r(\sqrt{2} - 1) \approx 0,414r$$ This is the radius of the largest atom that will fit into the site without deforming it.	***coordinates:*** **(1/4 1/4 1/4); (3/4 1/4 1/4); (1/4 3/4 1/4); (1/4 1/4 3/4) ; (3/4 3/4 1/4) ; (1/4 3/4 3/4) ; (3/4 1/4 3/4) ; (3/4 3/4 3/4).** The radius of the tetrahedral site is such that: The half-diagonal of the cube with edge a/2 is equal to the sum of the radius of the atom and the radius of the tetrahedral site: 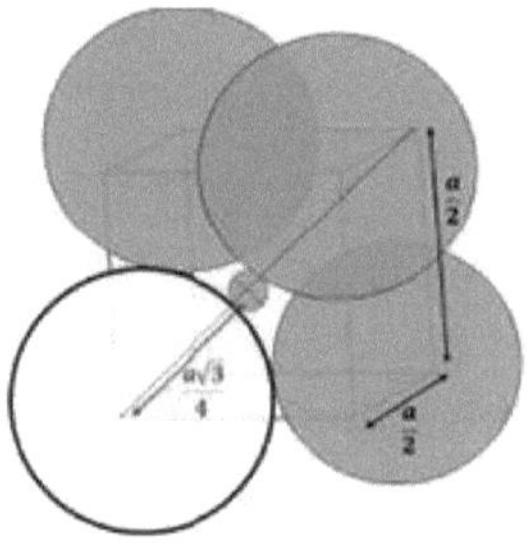 $$r + R_T = a\frac{\sqrt{3}}{4} \ avec \ 4r = a\sqrt{2}$$ $$R_T = r\left(\sqrt{\frac{3}{2}} - 1\right) \approx 0,225r$$

III. Compact hexagonal structure (HC) III.1 The mesh

Description of the HC mesh: a right prism with a rhombic base (one third of a hexagonal-based mesh), a node at each vertex (the sphere centred on such a node is divided between 8 meshes) and a node inside the mesh projecting, if the rhombic base is divided into two equilateral triangles, to the centre of one of these triangles.

The mesh is defined by the parameters $a = 2r$; $c = 4r\sqrt{\frac{2}{3}}$;$\alpha=\beta=90°$ et $\gamma=120°$.

Number of spheres (atoms) per mesh: 8x(1/8) + 1 = 2 or 4x(1/12)+4 x(1/6) + 1 =2.

Number of form groups Z=1.

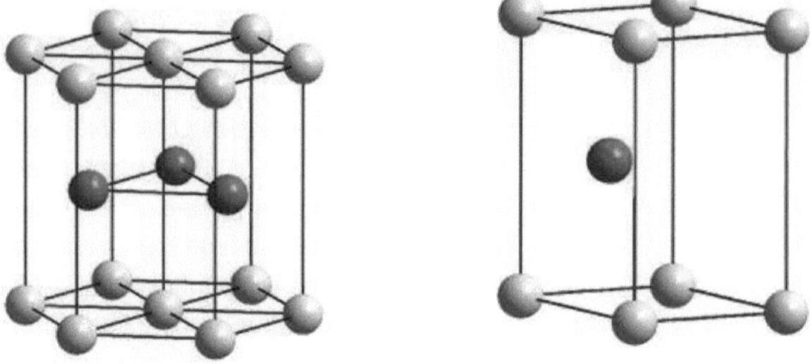

Conventional HC mesh in perspective. HC elementary mesh in perspective.

III.1. 2. Coordination

Each sphere is in contact with **12** other spheres: 6 from the same plane, 3 from the previous plane and 3 from the next plane. The co-ordination (or co-ordination index) is said to be 12.

III.2. 3. Direction of sphere tangency and the relationship between mesh dimensions and atom size

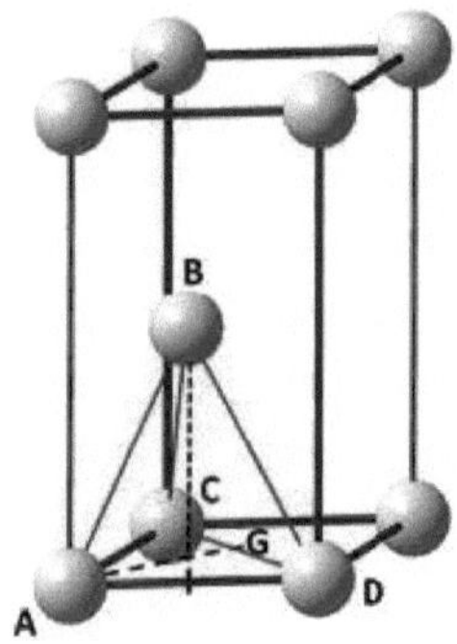

By construction, the spheres are tangent along the edges of the rhombuses in plane A. The relation $a=2r$ follows. In addition, there is tangency between spheres on two successive planes A and B, which imposes a condition between a and c. To obtain this, let's use the geometric properties of the ***tetrahedron ABCD***. In the basic triangle, ***the height*** of the T3T ***equilateral triangle ACD*** of side a is worth $AG = a\frac{\sqrt{3}}{2}$. The centre of gravity ***I*** of triangle ***ACD*** is located at *2/3* of the height ***AG***, i.e. at the distance from the vertex $AI = \frac{a}{\sqrt{3}}$. In T3 writing that the triangle ***ABI*** is right-angled at ***I***. Pythagoras gives: ($AB^2 = AI^2 + BI)^2$

➔ $BI^2=(c/2)^2=AB^2-AI^2=a^2-(\frac{a}{\sqrt{3}})^2$ We find the relationship

: $\frac{c}{a}=\sqrt{\frac{8}{3}}=2\sqrt{\frac{2}{3}}$ *or a= 2r* ➔ $c=4r\sqrt{\frac{2}{3}}\approx 1,63$

III.4. Compaction and density

Let's calculate the compactness of the system in the prismatic mesh. The base surface is
that of the rhombus with dimension *a,* so the area of the base surface is a^2 sin(2n/3).

The volume of the prismatic mesh with a diamond base is therefore $V=a^2.c.\frac{\sqrt{3}}{2}$. We obtain the value of the compactness *C* by dividing the volume of the two atoms belonging to the mesh by the volume of the mesh. Taking into account the previous results, we write, all calculations done:

$$C=C=\pi\frac{\sqrt{2}}{6}=0,74$$

Of course, we had to find the same compaction value as in the *cfc* structure, given the similarity of the stacking modes. We can also calculate the density as before:

$$\rho=\frac{ZM}{N_A V_h}\text{ avec } Z=2 \text{ et } V_h=a^2.c.\frac{\sqrt{3}}{2}$$

IV. Pseudo-compact stack

Mesh description: cubic, a node at each vertex (the sphere centred on this iHL'ud is divided between 8 meshes) and a node at the centre.
Number of spheres (atoms) per mesh: 8x(1/8) + 1 = 2.

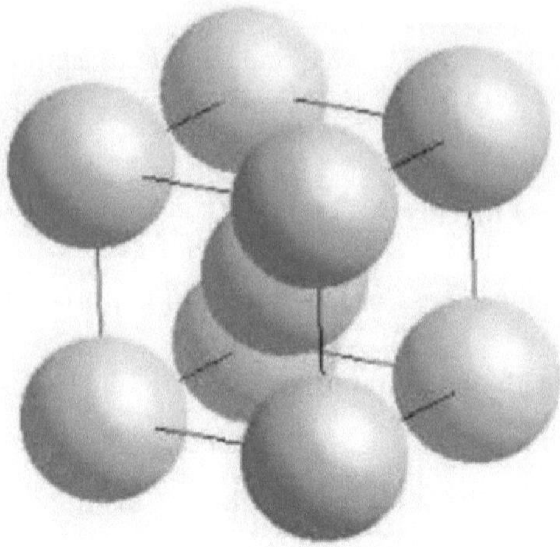

Compact cubic centric mesh. [2].

Each sphere is surrounded by eight equidistant spheres: co-ordination or co-ordination index = 8 (lower than that of the cfc and hc structures). The spheres are tangent in the direction of the cube diagonal, hence the relationship between the metal radius r and the mesh edge a.

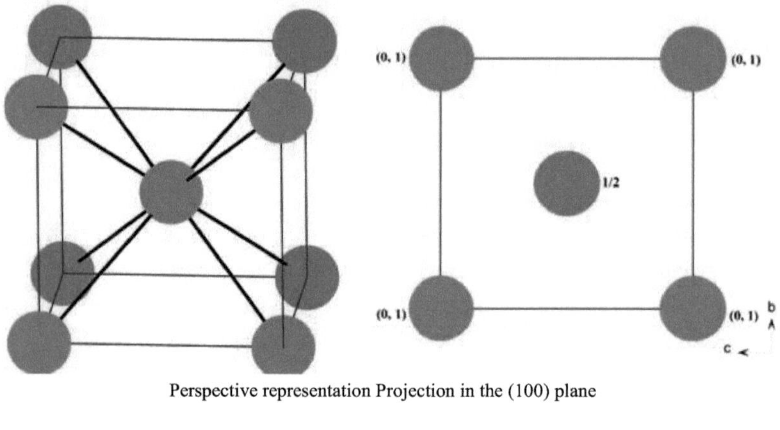

Perspective representation Projection in the (100) plane

The compactness of CC is expressed by :

$$C = \frac{Z.\frac{4\pi r^3}{3}}{a^3} \text{ avec } Z=2 \text{ et } a\sqrt{3} = 4r$$

$$C = \frac{\pi\sqrt{3}}{8} = 0,68$$

Compaction = 68% (lower than cfc and hc structures - less sphere compaction).

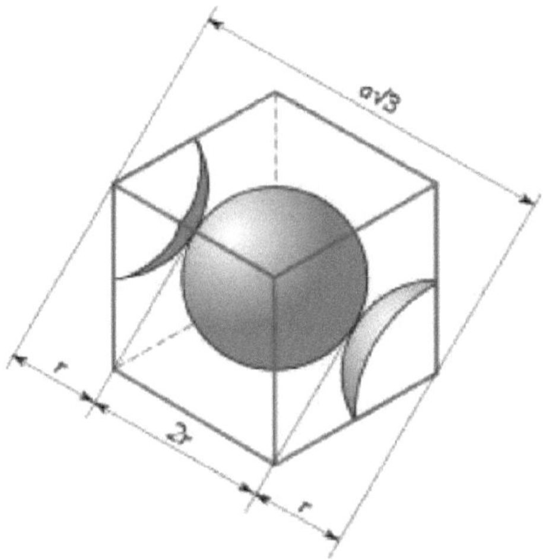

Relationship between the cubic mesh parameter a and the radius r of the metal particle in a centred cubic structure.

cubic structure. [1]

V. Alloys

Are systems formed by two or more metals or by a metal and a non-metal. They are also called solid solutions. They are prepared by cooling a liquid metal phase. There are two main families of alloys:

V.1 Substitution solid solution

These alloys are generally obtained from metals of similar size and with the same structural type. If ***M*** and ***M' are*** two metals of comparable radius, ***M'*** can be substituted for ***M*** in the lattice, and the general formula of the alloy is $M_{1-x}M'_x$. Substitution can be total or partial.

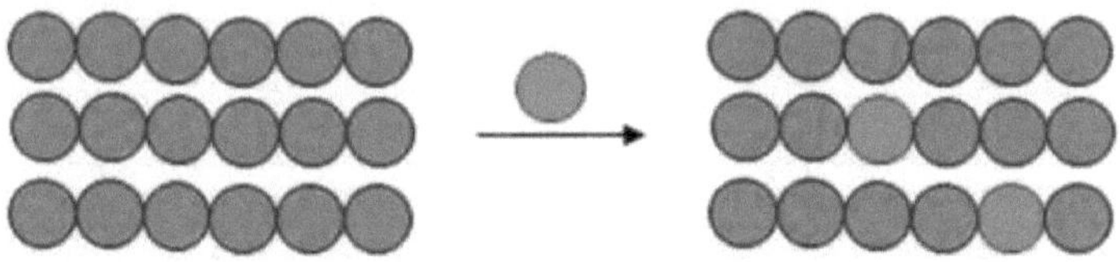

Example: Au-Cu (jewellery); Cu-Sn (bronzes); Cu-Al; Fe-Al.

V.2 Solid insertion solution

These solid solutions appear when one of the atoms, generally non-metal ***X****,* is significantly smaller than that of the metal. The non-metal then occupies the sites defined by the metal lattice, without causing any noticeable deformation. Consider a compact stack of atoms ***M***, of multiplicity. There are tetrahedral or octahedral interstices in which ***X*** atoms can be accommodated without distorting the lattice. The general formula for such an alloy is MX_x.

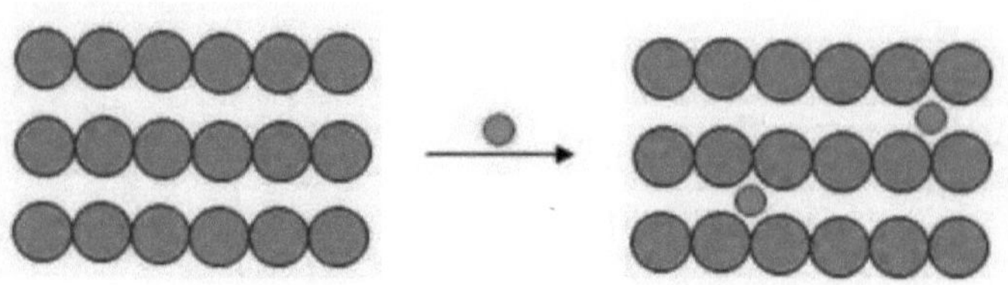

Example: Carbides: the carbon is inserted into the metal lattice: FeC3 (Cementite); TiC; CrC. Hydrides: hydrogen is inserted into the metal lattice: La, Pd, Pt, Zr, etc.
Nitrides: the nitrogen is inserted. FeN.

References :

[1] http://deuns.chez.com/sciences/cristallo/46.gif.
[2] https://ressources.univ-lemans.fr/AccesLibre/UM/Pedago/chimie/01/06-Etat_solide/deug/reseau.html.

CHAPTER 3

Ionic crystals

CHAPTER 3

Ionic crystals

I. Cohesion of the ionic crystal

The ionic crystal is made up of anions and cations. These oppositely charged ions maintain electrostatic bonds that hold the crystalline structure together. The larger anions occupy the mends of a host network, while the smaller cations are housed in certain sites of this host network. As the overall structure must remain neutral, the number of cations per cell must be identical to the number of anions. In addition, the electrostatic forces between the ions tend to repel ions of the same sign and attract ions with opposite sign charges. This leads to the following average equilibrium situation:

- there is contact between ions with opposite charges;
- there can be no contact between ions of the same sign.

We are going to study two types of ionic crystals: Type AB: CsCl; NaCl and ZnS (blende) and type AB2: CaF_2.

II. AB or MX type structure

II.1. . CsCl type structure

a) Mesh size

Chloride ions Cl^- form a simple cubic lattice by occupying the vertices of a cube of arete a. Cesium ions Cs^+ occupy the centres of the cubes. By tracing the neighbouring meshes we see that symmetrically the cesium ions form a cube of the same a-edge at the centre of which is a chloride ion. We give: $r(Cs^+) = r =$ 167 pm ; $r(Cl^-) = R = 181$ pm ;

- Conventional mesh parameter :

The Cs and Cl ions are tangent along the long diagonal of the cube, i.e. $a\sqrt{3} = R + 2r + R$

2

d or: $a = \frac{2}{\sqrt{3}}(r + R) = 404$ pm.

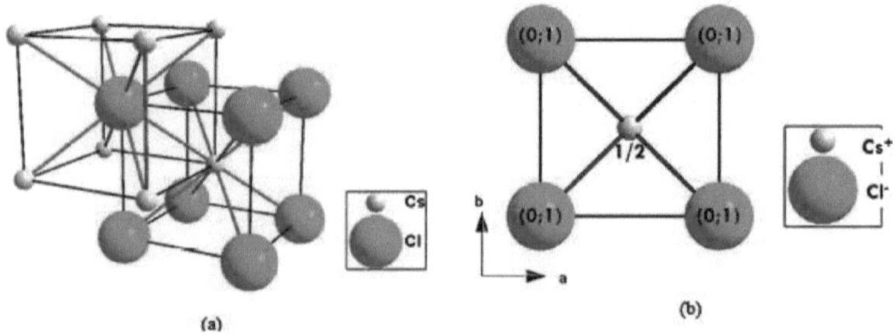

(a) Cl- and Cs+-based elementary mesh of the CsCl structure; (b) Projection of an elementary mesh of the CsCl structure in the (ab) plane.

- Number of ions per mesh and compactness :

The cesium ion is internal to the mesh and therefore counts as 1. The 8 chloride ions each participate in 8 different cubes and therefore count as 1/8.

This gives us 1 Cs ion$^+$ and 1 Cl ion$^-$ per cell. The crystal is therefore neutral.

Compaction is then :
$$C=\frac{\frac{4}{3}\pi(r^+)^3+\frac{4}{3}\pi(r^-)^3}{a^3}=0{,}683$$

- Coordinates:

Co-ordination of the anion with respect to the cation (co-ordination of the cation with respect to the anion) = Cl /Cs^{-+} =8/8: cubic site.

Co-ordination of the anion with respect to the anion (co-ordination of the cation with respect to the cation) = Cl /Cl^{--} = Cs /Cs^{++} =[6]

- Reduced coordinates

Origin on the anion :

Cl$^-$: (0 0 0)

Cs$^+$: (1/2 1/2 1/2)

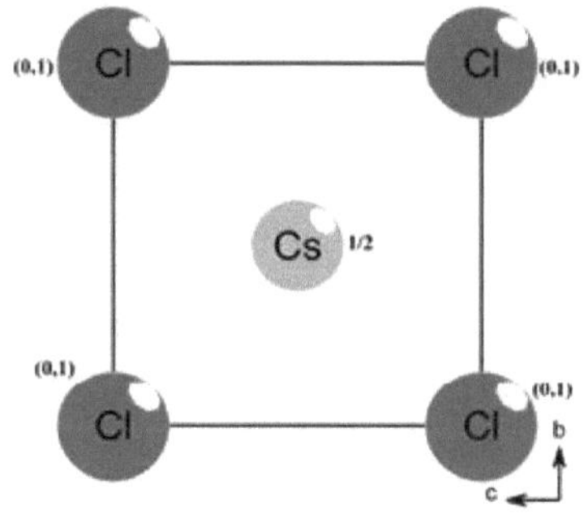

Projection of the CsCl mesh onto the (bc) plane

This structure can be described by two simple cubic arrays of Cl- anions and Cs+ cations offset from each other by ^ along the diagonal of the cube, i.e. by a translation (1/2 1/2 1/2).

b) Stability condition

The adjacent Cl$^-$ ions, which are larger than the Cs$^+$ ions, must not interpenetrate. The shortest distance **a** between two Cl ions$^-$ must therefore be greater than 2r and in the limiting case the anions are tangent to each other, i.e. :

$\mathbf{a \geq 2r^-}$ (1)

The Cs^{+} cation is placed in the interstices left free by the anions, assuming anion-cation contact:

$$2r^{+} + 2r^{-} = a\sqrt{3} \quad (2)$$

or $$a = \frac{2(r^{+} + r^{-})}{\sqrt{3}} \quad (3)$$

Replacement (3) in (1) $\rightarrow \frac{2(r^{+} + r^{-})}{\sqrt{3}} \geq 2r^{-} \quad (4)$

So $1 + \frac{r^{+}}{r^{-}} \geq \sqrt{3}$ or $\frac{r^{+}}{r^{-}} \geq \sqrt{3} - 1 \quad (5)$

As $r^{+} < r^{-}$ d'ou $\frac{r^{+}}{r^{-}} < 1 \ (6)$

(5) et (6) $\rightarrow \quad \sqrt{3} - 1 \leq \frac{r^{+}}{r^{-}} < 1$ ou $0,732 \leq \frac{r^{+}}{r^{-}} < 1 \ (7)$

5.1.. NaCl-type structure a) The mesh

The chloride ions Cl form a face-centred cubic lattice with an arete a.

The sodium ions Na occupy the octahedral sites of the c.f.c. network.

By tracing the neighbouring meshes we can see that symmetrically the sodium ions form a cfc network with the same end point at ; the chloride ions occupy the octahedral sites of this network.

We give: $r(Na^{+}) = r = 102$ pm; $r(Cl^{-}) = R = 181$ pm; [1].

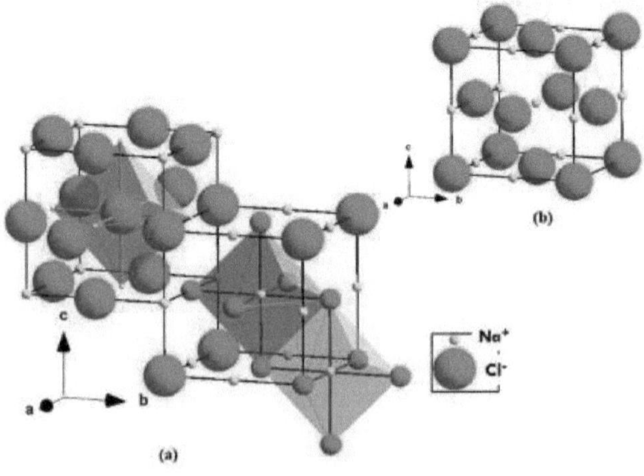

(a) NaCl crystal structure; (b) NaCl elemental lattice.

- Conventional mesh parameter :

The Na^+ and Cl^- ions are tangent along the edge of the cube, i.e. a = R +2r + R = 556 pm.

- Number of ions per mesh and compactness :

There are 4 octahedral sites per mesh, so there are 4 sodium ions per conventional mesh.

The 8 chloride ions on the vertices each participate in 8 different cubes and therefore count as 1/8. The 6 chloride ions on the faces participate in 2 different cubes and therefore count as 1/2. There are therefore 4 chloride ions per cell and the crystal is neutral.

Compaction is then : $C = \frac{4\frac{4}{3}\pi(r^+)^3 + 4\frac{4}{3}\pi(r^-)^3}{a^3} = 0{,}667$

Reduced coordinates

Origin on the anion :

Cl^- : (0 0 0) (1/2 1/2 0) (1/2 0 1/2) (0 1/2 1/2)

Na^+ : (1/2 0 0) (0 1/2 0) (0 0 1/2) (1/2 1/2 1/2)

- Coordinates:

Co-ordination of the anion with respect to the cation = co-ordination of the cation with respect to the anion = [6].

Coordination of l'anion with respect to l'anion = coordination of cation with respect to cation = [12].

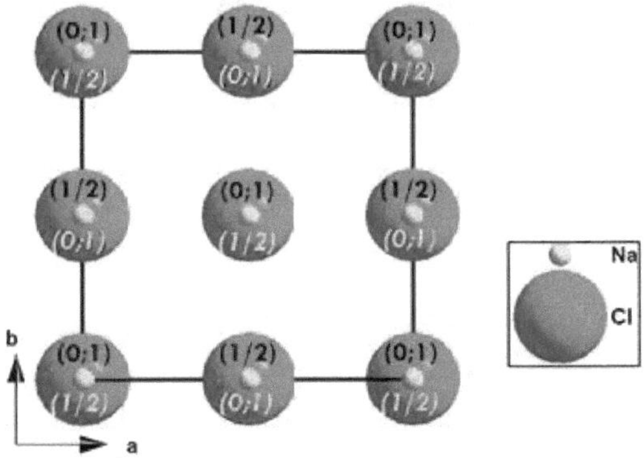

Projection of the NaCl mesh onto the (ab) plane.

This structure can be described by two cfc networks of Cl- anions and Na+ cations offset from each other by / along the edge or diagonal of the cube by either translation ((1/2 0 0) or (0 1/2 0) or (0 0 1/2)) or (1/2 1/2 1/2).

b) Stability condition

Considering one side of the cube, the boundary structure is obtained when the anions and cations are tangent along the edge of the cube, which corresponds to the relationship: $2r+ + 2r^- = a$ (8)

At the limit, the anions can only be tangent along the diagonal of the edge, i.e. :

$$2r^- \leq a\frac{\sqrt{2}}{2} \quad (9)$$

(8) et (9) → $2r^- \leq 2\,(r+ + r-)\frac{\sqrt{2}}{2}$ (10)

Soit $\frac{r^+}{r^-} \geq \sqrt{2}$ -1

Or

Given relation (11) and the boundary condition for stability of the CsCl type structure (relation 7), the boundary condition for stability of the NaCl type structure is :

$$\sqrt{2}-1 \leq \frac{r^+}{r^-} \leq \sqrt{3}-1$$

8) .3. ZnS type structure

a) The mesh

The sulphur ions S^{2-} form a cfc cubic lattice of arete a.

Zinc ions Zn^{2+} occupy half of the tetrahedral sites in the c.f.c. network.

Such a structure is called a blende-type structure. We give: $r(Zn^{2+}) = r = 74$ pm; $r(S^{2-}) = R = 184$ pm.

- Conventional mesh parameter :

The Zn^{2+} and S^{2-} ions are tangent along the long diagonal of the cube, i.e. for a quarter of this diagonal a $\frac{\sqrt{3}}{4} = r^+ + r^-$

d or : $a = \frac{4}{\sqrt{3}}(r^+ + r^-) = 404$ pm).

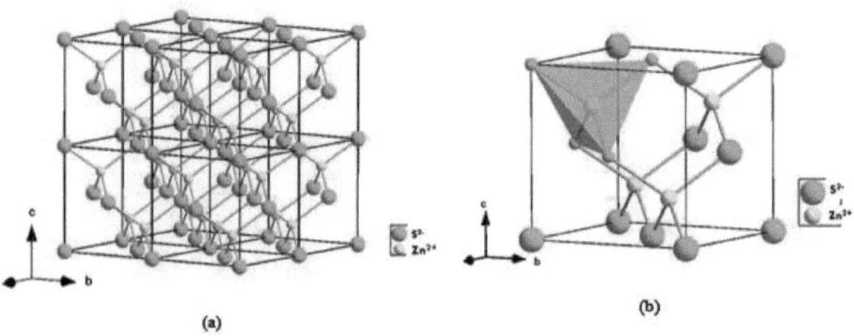

(a) Crystal structure of ZnS (blende); (b) Elemental lattice of ZnS (blende).

The experimental value is quite far from the theoretical predictions (a = 541 pm). By observing the electronegativities of the two elements sulphur and zinc, we can see that their difference is small, so that covalent bonds cannot be neglected. The ZnS crystal is therefore not a true ionic crystal but a hybrid between an ionic crystal and a covalent crystal.

- Reduced coordinates

Origin on the anion :

S^{2-} : (0 0 0) (1/2 1/2 0) (1/2 0 1/2) (0 1/2 1/2)

Zn^{2+} : (3/4 1/4 1/4) (1/4 3/4 1/4) (1/4 1/4 3/4) (3/4 3/4 3/4)

If we perform a translation of the type (1/4 1/4 1/4) we obtain the new coordinates

Origin on the cation :

S^{2-} : (1/4 1/4 1/4) (3/4 3/4 1/4) (3/4 1/4 3/4) (1/4 3/4 3/4)

Zn^{2+} : (0 1/2 1/2) (1/2 0 1/2) (1/2 1/2 0) (0 0 0)

Examination of these new coordinates shows that the Zn ions^{2+} also form a CFC network. The ZnS blende structure can therefore be described by two CFC networks, one formed by the S anions^{2-} , the other by the Zn cations^{2+} , offset from each other by 1/4 along the diagonal of the cube, i.e. by a translation of the type (1/4 1/4 1/4).

Number of ions per mesh :

There are 8 tetrahedral sites per mesh, so there are 8/2 = 4 Zn ions per conventional mesh.

The 8 S ions on the vertices each participate in 8 different cubes and therefore count as 1/8. The 6 sulphur ions on the faces participate in 2 different cubes and therefore count as 1/2. This gives us 4 S ions^{2-} per mesh and the neutrality of the crystal.

Compactness is then :

$$C=\frac{4\frac{4}{3}\pi(r^+)^3+4\frac{4}{3}\pi(r^-)^3}{a^3}$$

"Coordinences: Coordinence of the anion / the cation (coordination of the cation with respect to the anion) = [4].

Co-ordination of the anion / to the anion (co-ordination of the cation to the cation) = [12].

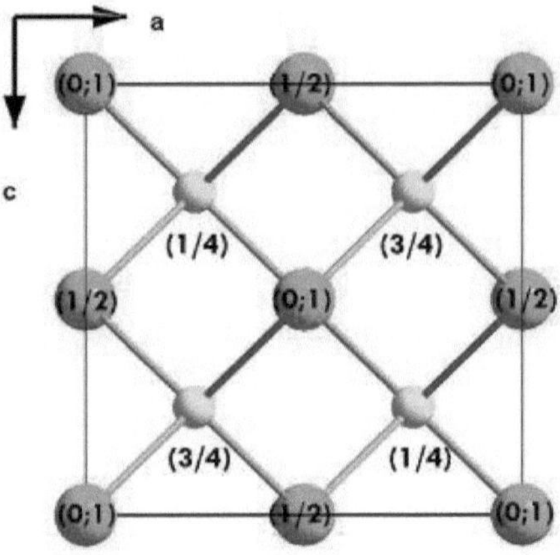

Projection of the ZnS mesh onto the (bc) plane

b) Stability boundary condition

The Zn^{2+} and S^{2-} ions are tangent along the diagonal of a small cube of arete a/2, so :

$$r^+ + r^- = a\frac{\sqrt{3}}{4} \quad (13)$$

Hence a = (14)

On the other hand, for the anions not to overlap on the diagonal of a face, :

$$2r^- \leq a\frac{\sqrt{2}}{2} \quad (15)$$

(14) et (15) $\rightarrow 2r^- \leq 4\frac{(r^+ + r^-)}{\sqrt{3}}\frac{\sqrt{2}}{2}$ (16) D'ou $\quad \frac{r^+}{r^-} \geq \frac{\sqrt{3}}{\sqrt{2}}-1$

Hence

II.4. Ion rays and structures

To a first approximation, the structural type adopted by an ionic compound consisting of an anion and a cation (MX) depends on the ratio of the ionic radii r /r^{+-} . The ratio of ionic radii determines coordination. Four main areas can be distinguished:

Structure		Condition of existence	Coordination
Type	Examples		Number Figure
CsCl	CsBr, CsI	$0.732<r^+/r^-<1$	[8]/[8] Cube
NaCl	MgO, CaS	$0.414<r^+/r^-<0.732$	[6]/[6] Octaedre
ZnS (blende)	BeO, MgTe	$0.225<r^+/r^-<0.414$	[4]/[4] Tetraedre

As the radius of the cation increases for a given anion, the ratio r /r^{+-} also increases. The larger the cation, the more anions it can surround itself with, and its co-ordination increases from 4 to 8.

111. Structure type AB2 / A2B or MX2/ M2X

111.1.CaF2 (Fluorine) type structure

Since the ionic radii of the Ca^{2+} and F- ions are: r(Ca^{2+}) =1.12A and r(F-)=1.31 A, the ratio r^+ / r- = 0.855 A falls within the stability range of the CsCl structural type (0.732 < r^+ / r^- < 1). However, if the CaF2 structure were of the CsCl type, the elemental lattice would contain an F $anion^-$ (at the apex of the lattice) and a Ca $cation^{2+}$ (at the centre of the lattice): electrical neutrality would therefore not be respected.

Since there are as many cubic sites as there are F- anions making up a simple cubic lattice, and since electrical neutrality implies the presence of twice as many fluorine atoms as calcium atoms, the occupancy rate of these sites by Ca^{2+} cations must be only 50%: Ca^{2+} cations would occupy half of the 8-coordination sites formed by F^- anions.

The calcium ions **Ca** form a c.f.c. cubic lattice of arete a.

Fluoride **F** ions occupy all the tetrahedral sites in the c.f.c. network.

- Number of ions per mesh :

There are 8 tetrahedral sites per mesh, i.e. 8 F^- per mesh and 4 Ca^{2+} per mesh.

- Reduced coordinates

Origin on the anion :

Ca^{2+} : (0 0 0) (1/2 1/2 0) (1/2 0 1/2) (0 1/2 1/2)

F^- : (3/4 1/4 1/4) (1/4 3/4 1/4) (1/4 1/4 3/4) (3/4 3/4 3/4) (1/4 1/4 1/4) (3/4 3/4 1/4) (3/4 1/4 3/4) (1/4 3/4 3/4)

- Coordinates:

Co-ordination of the anion with respect to the cation (co-ordination of the cation with respect to the anion) := F^- / Ca^{2+} = [4] and Ca^{2+} / F^- = [8].

Co-ordination of l'anion with respect to l'anion (co-ordination of cation with respect to cation) = [12].

Na_2O has an anti-fluorine structure in which cations and anions play the opposite roles to those in CaF_2.

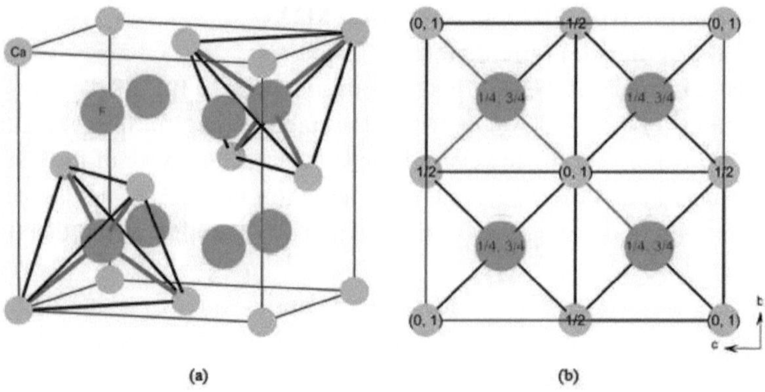

Perspective representation of an elementary mesh of the CaF2 structure; (b) Projection of the CaF2 mesh on plane (bc).

111.2.Na2O-type structure (anti-fluorine)

There is an anti-fluorine structure corresponding to solids of type M_2 X: it is derived from the fluorine structure by permutation of the positions of the anions and cations. This is the case with lithium, sodium and potassium oxides, sulphides, selenides and tellurides, as well as Cu_2 S, Cu_2 Se

O ions^{2-} replace Ca ions^{2+} forming a CFC lattice. The Na ions^{+} replace the F- ions occupying all the tetrahedral sites, i.e. the centre of all the small cubes of a/2 arete: they thus form a simple cubic lattice of a/2 mesh parametre.

The coordination of the Na cation^{+} is therefore 4, and that of the O anion^{2-} is 8: this is 4-8 coordination. The structure, with 8 Na cations^{+} and 4 O anions^{2-} per cell, therefore has 4 Na2O units per cell.

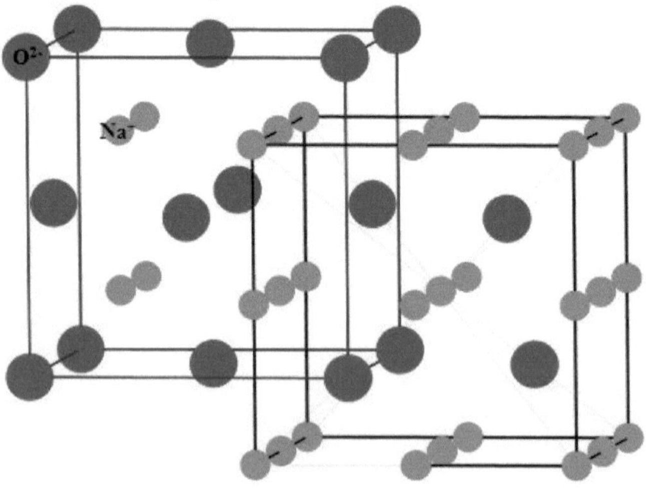

Crystal structure of Na2O.

References :

[1] Shannon, R. T., & Prewitt, C. T. (1970). Revised values of effective ionic radii. *Acta Crystallographica Section B: Structural Crystallography and Crystal Chemistry*, *26*(7), 1046-1048.

CHAPTER 4

Covalent crystals

CHAPTER 4

Covalent crystals

I. Cohesion of the covalent crystal

The atoms are linked by covalent bonds to form a one-, two- or three-dimensional macromolecule.

II. Diamond carbon

In this structure, the carbon atoms are hybrid sp^3 of the AX4 type, each of which is placed at the centre of a tetrahedron whose vertices are occupied by four other carbon atoms, which enter into four C-C covalent bonds with the reference atom. The shortest distance between two carbon atoms is **154 pm**.

The diamond structure results from the interlocking of two arrays of cfc dots offset by a quarter of a large diagonal. The pattern is then a carbon atom, located on an iHL'ud of each of the dot arrays. **The carbon atoms are stacked in a face-centred cubic pattern, occupying half of the tetrahedral sites**. There are therefore 4 + 4 = **8 carbon atoms per mesh**. **The coordination of** the carbon is equal to **4**.

This structure is similar to a blende-type structure. The crystal is shown in exploded view and in the (010) plane.

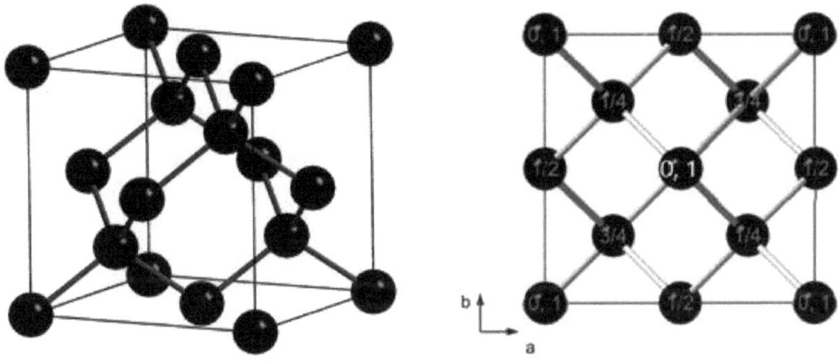

(a) (b)

(a) Tetragonal co-ordination in diamond; (b) Diamond crystal: projection on the (010) plane.

The compactness of a diamond is written :

$$C = \frac{8\times\frac{4}{3}\pi r_c{}^3}{a^3} = \mathbf{0,34\ avec\ (2r = a\sqrt{3}/4)}$$

The density is written as :

$$\rho = \frac{ZM_c}{N_A.a^3} = \frac{8\times 12}{6,022.10^{23}.(a(cm))^3} = (\text{g.cm}^{-3})$$

Diamonds are extremely hard. It also has exceptional brilliance.

due to the high value of its refractive index, linked to the nature of the covalent interaction between the atoms (very high melting temperature (3600°C); very high hardness; high density ($3.51 g/cm^3$). Nevertheless, diamond is relatively fragile. It is also an absolute electrical insulator. Silicon, on the other hand, which crystallises in a similar structure, is a semiconductor.

III. Carbon graphite

The graphitic carbon network is made up of regularly spaced and offset sheets forming a lamellar structure (fig.(a)).

In these sheets the carbon atoms are hybrid sp^2 of the AX3 type are placed at the vertices of regular hexagons with a dimension of 142 pm. Only 3 of the carbons in a ring in one plane project onto the carbons in the next plane, the other 3 project into the centre of the hexagon. Carbon coordination = 3: each carbon atom is surrounded by 3 atoms located in the same plane, the angle between two $(C\widehat{C}C)$ bonds is 120° (fig.(b)).

The planes are held together by Van der Waals interactions. These planes can slide over each other (lubricating properties). The non-hybrid electrons form a molecular orbital distributed over the entire sheet. This gives graphite high electrical conductivity in the plane of the sheet. Conduction in a direction perpendicular to the sheets is very low.

The projection of the graphitic carbon structure and an elementary mesh of its structure in the (ab) plane are shown in figures (a), (b) and (c).

The relationship between the mesh parameter ***a*** and the radius of the carbon atom r_c is explained in figure (d).

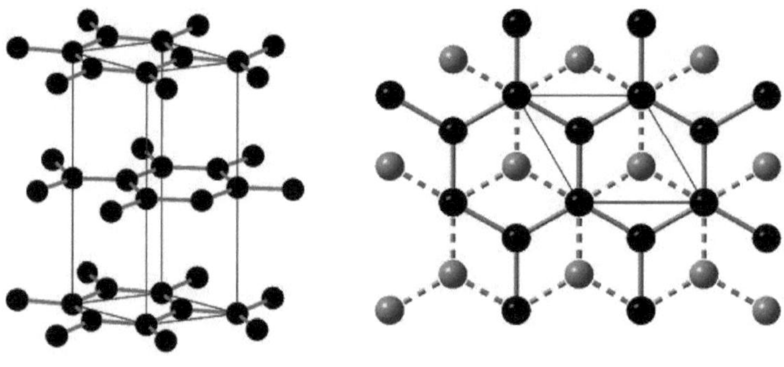

(a) (b)

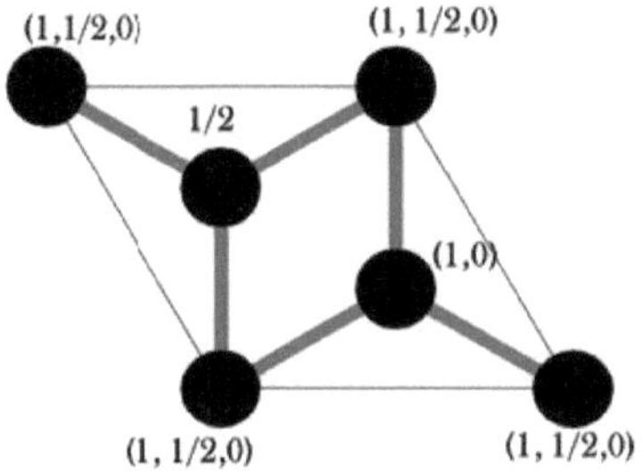

(c)

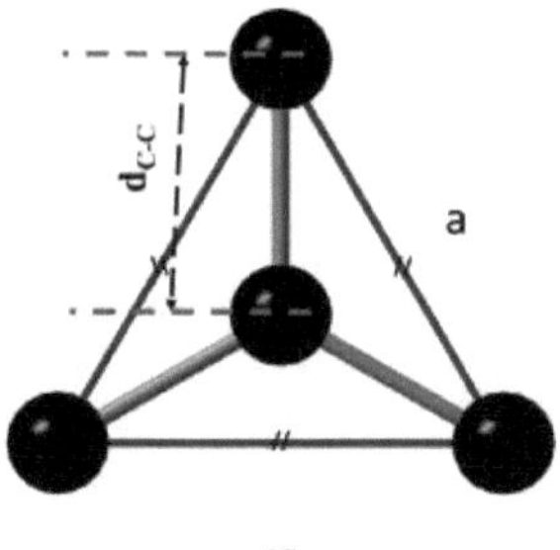

(d)

$dc - c = \frac{2}{3} a \times \frac{\sqrt{3}}{2}$

$\Rightarrow a = \sqrt{3} \times d_{c-c} = \sqrt{3} \times 2R_c$

- Mesh parameters: a = 1.42 A and c = 6.79 A - Multiplied^ of the mesh: m = 4x1/6 +4x1/12 +2x1/3 +2x1/6 + 2x1/2+ 1 = 4 atoms of earbone.

Atomic positions: (0 0 0) ; (2/31/3 0) ; (0 0 1/2) ; (1/3 2/31/2).

- Compactness

$$C = \frac{Z \times \frac{4}{3}\pi \times R_C^3}{a^2 \times c \times \sin(\frac{2\pi}{3})}$$

$$C = \frac{1 \times (4 \times \frac{4}{3}\pi \times (0,71)^3)}{(2,459)^2 \times 6,7 \times \frac{\sqrt{3}}{2}} = 0,17$$

Density :

$$\rho_{\text{graphite}} = \frac{Z \times M(C)}{N_A \times a^2 \times c \times \sin(\frac{2\pi}{3})}$$

$$= \frac{1 \times (4 \times 12)}{6,023 \times 10^{23} \times (2,459 \times 10^{-8})^2 \times 6,7 \times 10^{-8} \times \frac{\sqrt{3}}{2}}$$

$$= 2,2\ g.cm^{-3}$$

Molar volume :

$$V_M = \frac{V_{maille}}{quantité\ de\ matière} = \frac{(2,459 \times 10^{-10})^2 \times 6,7 \times 10^{-10} \times \frac{\sqrt{3}}{2}}{(\frac{4}{6,023 \times 10^{23}})}$$

$$= 5,28 \times 10^{-6}\ m^3.mol^{-1}.$$

Note

Graphite is the stable allotropic variety of carbon under standard conditions at

298 K. By heating under high pressure, graphite is transformed into diamond. The transformation takes place at around 2000 K, under 1500 bars. It supplies industrial diamonds.

MIX
Papier aus verantwortungsvollen Quellen
Paper from responsible sources
FSC® C105338

Printed by Books on Demand GmbH, Norderstedt / Germany